FRANZ BRENTANO:
Philosophical Investigations on Space, Time
and the Continuum

Franz Brentano

Philosophical Investigations on Space, Time
and the Continuum

Translated by Barry Smith

CROOM HELM
London ● New York ● Sydney

© 1988 Barry Smith
Croom Helm Ltd, Provident House, Burrell Row,
Beckenham, Kent, BR3 1AT
Croom Helm Australia, 44-50 Waterloo Road,
North Ryde, 2113, New South Wales

Published in the USA by
Croom Helm
in association with Methuen, Inc.
29 West 35th Street
New York, NY 10001

British Library Cataloguing in Publication Data

Brentano, Franz
    Philosophical investigations on space,
    time and the continuum.
    1. Space and time
    I. Title II. Philosophische Untersuchung
    zu Raum, Zeit und Kontinuum. *English*
    115   BD632
ISBN 0-7099-4476-4

Library of Congress Cataloguing-in-Publication Data

ISBN 0-7099-4476-4

CONTENTS

Editors' Introduction to the English Edition

Stephan Körner and Roderick M. Chisholm

Introduction

Analysis of the concepts of space and time and of the more general concept of a continuum is an essential part of natural philosophy and descriptive psychology. It was only natural, therefore, that Brentano occupied himself with these questions for the greater part of his life, from the time of his early efforts to set forth the Aristotelian philosophy to the final years of his life when, completely blind, he set forth his final views on what he called 'descriptive psychology.'

The essays in the present volume are a selection from Brentano's works on space, time and the continuum. None of them have previously been published in English translation. Brentano had entrusted Alfred Kastil and Oskar Kraus with preparing his unpublished works for publication. Kastil had intended to publish selections from the works on space, time and the continuum in a single volume with an analytic table of contents, introduction and notes, but he died before he was able to complete the work.[1]

In selecting the manuscripts for the present volume, we have emphasised those works that are primarily concerned with *philosophical* problems and that represent Brentano's final views. We have tried to avoid unnecessary repetitions.

The basic features of Brentano's theory may be found in the first selection ('On what is continuous'); further details are added in the second selection ('On the measure of what is continuous'). To understand this theory in its broad outlines, one should compare it both with the doctrine of Aristotle and with the classic mathematical theories of Cantor and Dedekind. The comparison with Aristotle is essential since, as almost always, Brentano begins with the views of Aristotle and then modifies them in far-reaching respects. Comparison with the mathematical theories is essential in order to exhibit the details of Brentano's view and to remove certain misunderstandings.

Aristotle's Theory of the Continuum

Aristotle's theory of the continuum rests upon the assumptions that all change is continuous and that one understands nature only to the extent that one understands the nature of change. According to Aristotle, continuous alteration of quality, of quantity and of position is given in perception and in intuition; hence, if continuity seems to involve contradiction, the fault must be in our description of continuity.

Aristotle thought it necessary, therefore, to refute Zeno's so-called paradoxes of motion, for these were supposed to support the view of Parmenides according to which all change is illusory. The supposed paradoxes are based upon three presuppositions: (i) every path is composed of an infinite number of points; (ii) any subsection of any path is also composed of an infinite number of points, so that any two subsections, no matter how different in length they may be, may be brought into a one-to-one correlation with each other; and (iii) if an object moves over a path, then, no matter how small the path may be, the object moves over an infinite number of points all of which are distant from each other, however small the distance may be. (No distinction is made between arbitrarily small finite distances and infinitesimal ones in the sense of Leibniz and his successors.) Of these three presuppositions, the first two are accepted in contemporary and classical mathematics; the third, however, is subject to conflicting interpretations. The three together have the consequence that, despite all appearances, a moving body needs as much time to cover any proper part of a given path as it needs to cover that path itself; for in both cases the body must run through the same (infinite) number of points each of which is distant from every other.

On the assumption that set-theory, as it is now understood, is consistent, we may say that the first two presuppositions do not lead to an antinomy. Aristotle, however, rejects all three presuppositions. He has a simple and very good reason for this: they do not describe the kind of continuum that is given in perception or intuition; and therefore they are not relevant to the theory of such continua. Aristotle, like Brentano after him, wanted

viii

to describe those phenomena of continuity that are given in perception.

It was clear to Aristotle that a continuum cannot consist of points. Any two unextended points are such that either they fall completely together or they are completely separated; in the first case, they yield only a single unextended point; and in the second case they are points that are separated by a gap.

According to the Aristotelian theory, any continuum — say, a continuous path, or a duration of time, or a motion — may be divided *ad infinitum* into other continua but not into what one might call 'discreta.' More exactly, paths may be divided into smaller paths but not into unextended points; durations may be divided into shorter durations but not into unextended moments; and a motion may be divided into shorter motions but not into unextended 'stations.'[2] But this does not mean that a continuous line, for example, cannot be divided at a point which constitutes the common border of the line-segments divided by it. A common border connecting two parts of a continuum must be distinguished from the subcontinua connected by it.

In order to explain the nature of the common border which unites two continua into one, Aristotle distinguishes various types of relations which things of the same sort can bear to each other. Things succeed each other in a definite order if nothing of the same sort lies between them; one thing follows immediately upon another, if the boundaries of the two things are in contact; and they are continuous if their boundaries are one.[3] The common parts, by means of which the borders of a continuum are connected with each other, exist only potentially, for they exist only when they connect the parts of a continuum; and the parts in their turn are similarly dependent as parts upon the existence of the continuum.

The Classic Mathematical Theory

According to the classic mathematical theory of Cantor and Dedekind, a continuum actually consists of noncontinuous simple units — of a nondenumerable infinity of such units. It is not assumed, however, that such units are found in perception or intuition. In working out his own theory of the continuum, Brentano gave particular consideration to Dedekind's theory of real numbers, but his conception of continuity is quite different from that of Dedekind.[4] Dedekind assumed, with Cantor and with their nonintuitionistic or nonconstructivist followers, that the natural numbers, 1, 2, 3, . . . , are given, not only as an infinite series in which each number follows immediately upon another, but also as a completely given totality.

No infinite totality or set may be compared by counting with other infinite or finite sets, but comparison is possible by means of one-to-one correlation. Thus two sets are equivalent if there is a one-to-one correlation between their elements. One set is smaller than another, if the first may be put into one-to-one correlation with a subset of the second and no subset of the second may be put into one-to-one correspondence with the first. No finite set is equivalent to any of its subsets, but every infinite set is equivalent to one of its subsets; the set of all whole numbers, for example, is equivalent to the set of all even whole numbers. The equivalence of a whole to one of its subsets is thus taken as an essential mark of an infinite set.

Assuming the existence of the totality of all natural numbers, Dedekind defines the totality of all fractions. He defines a fraction as an ordered pair (for example, 1/2 as the pair whose first number is 1 and whose second number is 2) which satisfies certain postulates. Together these postulates correspond to the usual rules governing the use of fractions. The place of natural numbers is taken by fractions with denominator 1 (for example, 5 becomes the fraction 5/1).

The totality of rational numbers is *dense* but not *continuous*. It is dense in the sense that, between any two rational numbers there is a third (and therefore infinitely many others). But it is not continuous, for it does not contain all 'real numbers' (thus it does not contain the square root of two, since there is no fraction having a square that is equal to 2). Dedekind preserves the missing real numbers by defining the concept of a 'cut' for the totality of rational numbers. A cut is a set of rational numbers which is such that (i) the set contains some but not all rational numbers, (ii) every rational number contained in the set is smaller than any rational number not contained in the set, and (iii) the set contains no largest rational number — no rational number greater than every other rational number contained in the set. It can be shown that the totality of all cuts corresponds to the totality of all irrational numbers (such numbers as the square root of two). It can also be shown that the set of all real numbers is greater than the set of all rational numbers. (For the set of all rational numbers is equivalent in magnitude to a subset of the set of real numbers; and the set of real numbers is not equivalent to any subset of the set of all rational numbers.) The continuous totality of real numbers is thus a 'larger' infinite set — an infinite set of higher order — than that merely dense totality of all rational numbers.

Aristotle and Brentano reject the concept of an infinite totality and hence the idea that one such totality can be 'larger' or of higher order than another. Therefore they do not recognise the distinction between density and continuity — a distinction which can hardly be called phenomenological. And the thought of ordering infinite sets with respect to their size is completely foreign to the phenomenological approach of Aristotle and Brentano. The classic mathematical theory may be thought of as providing a kind of idealisation of the phenomena of continuity; but it does not provide a description of what is given in perception.

Brentano's Conception of the Continuous

Brentano's theory resembles Aristotle's, both in its fundamental approach — which is to regard continuity as a perceptual phenomenon, rather than a mathematical construction — as well as in many details. The principal metaphysical difference concerns Aristotle's view according to which the parts of a whole are only potential things and not actual things. The principal epistemological and psychological differences between Aristotle and Brentano are based upon (i) Brentano's highly original and important theory of relations, (ii) his views about how we acquire the concept of a continuum and (iii) his conception of the relation between continua and their boundaries.

Brentano's attitude toward the mathematical theories of the continuum of Dedekind, Cantor and their successors wavers between rejecting them as indadequate and according them the status of fictions. The reasons for this may be that Brentano was inclined to think of mathematical and physical theories as descriptions of experience rather than idealisations; and if the theories were considered as descriptions, then they would be 'misrepresentations' of empirical phenomena. Once we recognise the distinction between the mathematical and the phenomenological conceptions of continua, it should be clear that there is no conflict at all between the theories of descriptive psychology (in the sense of Brentano) and the theories of pure mathematics (in the sense of Dedekind) or the theories of mathematical physics (in the sense of Einstein).[5]

We derive the concept of that which is continuous, according to Brentano, from sensible intuition (*Anschauung*). Indeed, he holds that 'all of our sensible intuitions present us with that which is continuous — this is true, not only of outer perception but also of that inner perception which always accompanies outer perception, and hence it is true, too, of the sensible intuitions presented by memory.'

In what way does sensible intuition present us with that which is continuous? Brentano suggests that there are three moments in our apprehension of the continuous. First, sensation presents us with objects having parts that coincide. Secondly, given such objects, we are able to abstract the concept of a *boundary* and then apprehend that such objects contain boundaries that coincide. And, thirdly, given this apprehension, we have all that we need in order to have the concept of a continuum.

What continuous objects, then, does sensation present us with?

In visual sensation, we are presented with that which has length and breadth; hence, we are presented with an object which is such that between any two of its parts which are separated from each other there is a third part. In every sensation, we are presented with something having a certain degree of qualitative density; we are presented with a quality which is such that (i) the object could have that quality in a greater or lesser degree, and (ii) between any two degrees of that quality there is still another degree of that quality. We may also be presented with a qualitative continuum when, say, we have a visual sensation wherein there is a continuous increase from left to right in the quality of a certain colour. And, finally, every sensation presents us with what is temporally continuous. This presentation is especially striking when we apprehend something as moving or as at rest.

There are, therefore, *secondary* and *primary* continua. A secondary continuum, unlike a primary continuum, is one that is founded upon *another* continuum. Every continuum is founded upon a temporal continuum; thus whatever is spatial is also temporal. And every qualitative continuum is founded upon a spatial continuum.

Whenever we apprehend a continuous object, according to Brentano, we are able to apprehend with complete certainty that the whole contains countless boundaries and coincidences of boundaries. In saying that the intuited whole contains 'countless' boundaries and coincidences of boundaries, Brentano is not telling us that the whole contains an *infinite* number of such boundaries and coincidences of boundaries. For, according to him, the concept of an actual infinite leads to contradiction and therefore there cannot be said to be an infinite number of things. 'One can say of a continuum, then, only that it can be described as being as large a finite number of actual entities as you please, but not as an infinitely large number of actual entities. One may maintain only that there are potentially infinitely many entities.'[6]

Brentano's Theory of Relations

Brentano's theory of relations is important both to his theory of continuity and to his theory of intentionality. The theory is itself intentional in that it is derived from his conception of the ways in which we *relate* things — the ways in which we *think* about things relationally.

According to Brentano's reism, the object of thought or of any intentional attitude is always a concretum or individual thing. If we think of John being taller than Charles, then, according to Brentano, we are thinking of John as having the attribute of being taller than Charles. We are thinking directly of John and only indirectly of Charles; we are thinking of Charles only to the extent that we are attributing to *John* the attribute of being taller than Charles. John is the *fundamentum* of the relation and, Brentano says, is thought of *in modo recto*. Charles is the *terminus* of the relation and is thought of only in *modo obliquo* — only indirectly by way of the attribute that is attributed to John. (But if, now, we should think of Charles as being shorter than John, then Charles becomes the fundamentum of the relation and is thought of *modo recto* and John is the terminus and is thought of only in *modo obliquo*.)

In other words, to acknowledge the existence of a relativum is to acknowledge the existence of the fundamentum as characterised by a relational attribute. On this analysis the existence of the fundamentum and the existence of the terminus may, but need not, coincide. The necessity, impossibility or possibility of the coexistence of the fundamentum and terminus of a relation, are important criteria for distinguishing various types of relativa.

According to the usual theory of relations, the affirmation of any relation between two things — for example, John being taller than Charles — involves the application of the relation ('x is taller than y') to two terms (John and Charles, in that order) and therefore implies the existence of the two terms. But according to Brentano, this account holds only for some types of relation. He therefore looks separately at each term of a relation and at the way in which one is aware of that term. His theory is thus not so much a theory of relations, as of *relativa*, of things *qua* bearing relations to other things.

He considers the following types of relativum: (1) intentional; (2) comparative; (3) causal; and (4) temporal.

(1) In the case of *intentional* relativa, the existence of the fundamentum does not ordinarily imply either the existence or the non-existence of the terminus. From the fact that one thinks of a unicorn or of a golden mountain, it does not follow that there *is* a unicorn or a golden mountain, and it does not follow that there is *no* unicorn or golden mountain.

But there are certain intentional relations which are exceptions to this principle. One type of exception is the case in which the thinker affirms with evidence the object of his thought: here the existence of the fundamentum implies the existence of the terminus. Another type of exception is the case in which the thinker denies with evidence the object of his thought: here the existence of the fundament implies the non-existence of the terminus.

But evident acceptance and evident rejection *presuppose* thinking, and thinking need not involve the existence of its object. We could say, then, that one mark of an intentional relativum is this: it presupposes a fundamentum which need not involve the existence or non-existence of its terminus.

(2) *Comparative* relativa are similar in one important respect to intentional relativa. The relation may obtain even though the terminus does not exist. We may say that a horse looks like a unicorn without thereby committing ourselves to the existence of a unicorn. But there are important respects in which intentional relations differ from comparative relations.

Intentional relations cannot begin to obtain or cease to obtain until there is a real change in the fundamentum of the relation. This is not true of relations of comparison. Socrates can become taller than Caius, or cease to be taller than Caius, without Socrates himself changing in any way; for possibly it is only Caius who alters with respect to size. But Socrates cannot become, or cease to be, a fundamentum in an intentional relation unless *he* changes in a certain way — either by ceasing to think in ways he had been thinking or by thinking in ways he hadn't been thinking. Hence we could say that a second mark of an intentional relativum is this: it cannot come into being or pass away without there being a real change in the substance to which that relation is attributed.

(3) *Causal* relativa differ from intentional and comparative relativa in interesting ways. If we think of a cause as being *a thing that brings about a thing* and of an effect as *a thing that is brought about by something*, then, in those cases where the cause precedes the effect in time, the cause exists only when the effect does not exist.

(4) Finally, in *temporal* relativa consisting in a present event being preceded by a past event, the existence of the fundamentum implies the non-existence of the terminus since the preceding, past event must have ceased to exist. (Unlike many contemporary philosophers, Brentano takes tense seriously. What exists coincides precisely with what exists *now*.) In relativa which are wholes consisting of parts, the existence of the whole implies the existence of the parts, whereas the existence of a part need not imply the existence of the whole. (The existence of a wood implies the existence of all the trees, whereas the existence of one of them — say after a wood fire — is compatible with the non-existence of the wood.)

It is characteristic of intentional relativa, but not only of intentional relativa, that the fundamentum can exist even though their termini do not exist.

Boundaries as Relativa

In the case of a boundary and its continuum, the existence of the fundamentum implies the existence of the terminus and the existence of the terminus implies the existence of the fundamentum. The boundary of a continuum, therefore, differs from the fundamenta of all the other relativa considered by Aristotle and Brentano in existing *only* as a boundary. For example, a thinker is not dependent upon his having any thought; but a boundary that bounds a particular continuum *is* dependent for its being upon its being the boundary of *some part or other* of that particular continuum.

The boldness of Brentano's analysis becomes apparent if we consider this account of the nature of a boundary in relation to his distinction between spatial and temporal continua. Whereas a spatial continuum exists as a whole, a temporal continuum exists only *qua* boundary. What exists, according to Brentano, is coextensive with what exists *now*; but the now is a boundary between the past and the future; hence a temporal continuum exists only within this boundary. Yet the boundary would not exist if it were not the boundary of a temporal continuum. And the continuum which it bounds determines the nature of that boundary completely. Thus Brentano says both (i) that the temporal continuum exists only to the extent that its boundary exists and also (ii) that the temporal continuum is a *sine qua non* of the boundary. But the second of these two theses must be understood in terms of the doctrine of relativa. The boundary would not exist unless it were a relativum bounding either the past or the future or both.[7]

If one considers Brentano's theory of continuity in the context of his general theory of relativa, it appears rather different from the theory of Aristotle, whose central notion of a continuous boundary is more of a metaphor than the epistemologically analysed and ontologically clearly characterised concept of Brentano's 'Grenze.' The originality of Brentano's theory, as compared with Aristotle's of which it is a development, shows itself particularly in his classification of continua and in the bearing of his analysis of continuity on his theory of space and time.

Plerosis and the Coincidence of Boundaries

Brentano writes in the first essay presented here: 'That which characterises a continuum above all else is the thought of a boundary in the strict sense, which thought connects itself with the possibility of a coincidence of boundaries.'

The *plerosis* — or *fullness* — of a boundary is a function of the number of directions in which it is a boundary. Thus a boundary within a temporal continuum may be a boundary in one direction (if it is merely the endpoint of something that is past or if it is merely the beginning point of something that is future). Or it may be a boundary in two directions (if it is both an endpoint and a beginning point). But the boundary of a spatial continuum is not thus restricted with respect to the number of directions in which it may be a boundary. It may be a boundary in all the directions in which it is capable of being a boundary, or it may be a boundary in only some of these directions. If a boundary — whether temporal or spatial — is a boundary in all the directions in which it is capable of being a boundary, then it exists in 'full plerosis'; otherwise it exists only in 'partial plerosis.' Brentano writes:

> ... the spatial nature of a point differs according to whether it serves as a limit in all or only in some directions. Thus a point located inside a physical thing serves as a limit in all directions, but a point on a surface or an edge or a vertex serves as a limit only in some direction. And the point in a vertex will differ in accordance with the directions of the edges that meet at the vertex ... I call these specific differences distinctions of plerosis. Like any manifold variation, plerosis admits of a more and a less. The plerosis of the centre of a cone is more complete than that of a point on its surface; the plerosis of a point on its surface is more complete than that of a point on its edge, or that of its vertex. Even the plerosis of the vertex is the more complete the less the cone is pointed.[8]

Brentano believes that by means of the concept of plerosis he can speak in a certain sense of the 'parts' of a boundary even though the boundary may have no dimensions. It is one thing to speak of the present as being the end of the past and another thing to speak of the present as being the beginning of the future.

Consider now a traditional philosophical problem. 'If a thing begins to move, is there a last moment of its being at rest or a first moment of its being in motion? There cannot be both, for if there were, then there would be a time between the two moments, and at that time the thing could be said neither to be at rest nor to be in motion.'

The statement of the problem, Brentano would say, correctly recognises the *impossibility of adjacent points*. But what it fails to recognise is the *possibility of coincident points*.[9] Brentano's solution is to say that at one and the same moment the thing ceases to be at rest and begins to be in motion. The temporal boundary of the thing's being at rest (the end of its being at rest) is the same as the temporal boundary of the thing's being in motion (the beginning of its being in motion), but the boundary is twofold with respect to its plerosis. The boundary is in half plerosis at rest and in half plerosis in motion.

The Source of our Concept of Time

The source of our concept of time, according to Brentano, is the intuitive experience he calls 'proteraesthesis' or 'original association.' This experience, he insists, is to be distinguished from sensation. But it is a phenomenon that accompanies every sensation. Examples are the hearing of a melody, the seeing of something in motion, and the seeing of something at rest. In each case, we experience a succession: in the first case, we experience one note preceding another note; in the second case the moving object being now in one place and now in another; and in the third case one and the same thing remaining exactly where it was. The experience of any such succession involves what may be called an experience of the past. The duration of such proteraesthesis is very brief. For example, in a single experience we 'see' part of the circular motion of the second-hand of a clock, but we do not see the entire circular motion and if the motion were not sufficiently swift we would not see it at all. Yet brief as such experiences are, they enable us to acquire the concepts of past, present, and future, the concepts of before and after, and the concept of a temporal continuum extending indefinitely in two directions.

Consider now the proteraesthesis involved in the hearing of the first four notes of a melody, say, a, b, c, and d. This provides us with a paradigm case of the experience of a succession. The experience is not adequately described by saying, 'We experience a and then we experience b and then we experience c.' For such a statement would be a report only of a *succession of experiences*, not a report of an *experience of succession*.

Many have held that the field of consciousness is temporally extended in the way in which, say, the visual field may be said to be spatially extended. According to this view, just as a red spot can be at the left side of the visual field and a blue spot at the right, the note c can be in the present part of the auditory field while the note b is past and the note c in a part that is even more past. But does it make sense to say of the note b that it *is* past? If b is no longer in the present, we cannot say that it *is* in a part that is past. If we take tense seriously, as Brentano does, we cannot say of the field of consciousness or of the objects of sensation that they (now) *have* a temporal extension.

Upon hearing the first three notes of the melody, we are aware of all three in a single experience. Yet the experience of a succession is not the experience we have when we are aware of three notes sounding together as in hearing a chord. The three notes a, b, and c, if heard all at once, would provide a discord and not the beginning of a melody.

It is tempting, but hardly adequate to the facts of the matter, to describe the experience of proteraesthesis by appeal to memory. Thus one might say: 'First we hear the a, then we hear the b and at the same time remember having heard the a, and then, after that, we hear the c and remember having heard the b.' Realising that such a description still seems to leave something out, one may add: 'On hearing the c, we remember not only having heard the b but also having heard the b *while* remembering having heard the a.' The experience of the melody would thus be described as a combination of a present perception with the memory of a past perception and with the memory of another perception that is even more past. But this is hardly successful as a description of the experience of hearing a melody. The conclusion may seem inescapable, therefore, that we experience c as preceded by a and by b, and hence that, in experiencing c, we experience both a and b as past. But if we experience a and b *now*, then how can we be said to be experiencing them as past and therefore *not* now and as such that one of them is *before* the other?

We cannot say, of the earlier notes in the experience of proteraesthesis, that 'they exist in the present with the attribute of being past.' For nothing *has* the attribute of being past. If anything *has* a given attribute, then that thing exists now and cannot be said merely to exist in the past. How, then, are we using the word 'past' when we say that the note is presented as past?[10]

Brentano's final view is the following. Temporal differences within experience are to be thought of, not as differences in the objects that we are conscious of, but as differences in the ways in which we are conscious of those objects.[11] In short, there are temporal modes of presentation or thinking. It is one thing for a note, say, to be presented as present and another thing for it to be presented as past. (We should remind ourselves that, according to Brentano, a thing need not exist in order to be an object of presentation.) And one note may be presented as being *more* past than another.

In the third essay of Part Two of this book, Brentano summarises his final conception as follows: 'When something that had been given as present appears as more and more past, it is not the case that *different objects* are apprehended as existing; rather the same object is apprehended throughout, but in *different ways*, with a different mode of apprehension.'[12]

Brentano also held that the objects of the temporal modes of presentation — those modes other than the present — are always experienced in *modo obliquo*. According to his final view of temporal thinking, if we think of a certain thing as past or as future, then, actually, we are thinking of the thing in *modo obliquo* and we are thinking in *modo recto* of some present thing. This present thing could be oneself.

The Temporal and the Spatial

A theory of our consciousness of time is not as such a theory about the nature of time. Brentano's theory of our *consciousness* of time is a theory about the nature of our experience and about the modes or ways in which we apprehend things and make judgements about them. But his theory about the nature of *time* — more accurately, his theory about that which is temporal — is an ontological theory, in the strictest sense of the term 'ontology.'

To understand Brentano's theory of time, we must view it in terms of two of his most fundamental doctrines. The first is his reism: the doctrine according to which the only things that there are, in the strict and proper sense of 'there are,' are individual things or concreta or, as Brentano calls them, '*entia realia*.' The second is the doctrine, according to which the only things that there are, are things that exist at the present time, things that exist *now*.

Given the first of these two ontological doctrines, we cannot say that, in addition to particular individual things, there *is* a time or space in which these individual things are to be found.

Given the second of the two doctrines, we cannot say that things are *temporally* extended in the sense in which we may say that they are *spatially* extended. For the now is not temporally extended but exists only as a boundary between the past and the future. Or, to speak more accurately (for 'the now' does not refer to an *ens reale*), *things that exist now* — in short, *things that exist* — exist only as boundaries of what did exist or of what will exist or of both. And we cannot say that things stand in temporal relations to each other in any way that is analogous to that in which we can say that things stand in spatial relations to each other.

Thus Brentano writes: 'There is not a space and a time; rather, there is that which is spatial and that which is temporal.' The temporal is that which is temporally continuous. It is a mistake, therefore, to suppose that there is only *one* temporal continuum. As Kastil put it, 'there are as many temporal continua as there are things.' Hence we should not say that things are 'in time,' much less that they are '*in the same time*'; but we may say that all things are 'co-temporal.' That which is temporal coincides precisely with what there is. And what there is coincides precisely with what there is *now*. Everything that there is, then, exists as a boundary wherein past and future come together. Even God, according to Brentano, is a temporal being. The source of all change, Brentano says, cannot be itself a changeless being. In the sense in which anything may be said to be 'in time,' God is 'in time'; that is to say, he exists now.

Just as the temporal is to be identified with that which is real or individual as such, so the spatial is to be identified with that which is corporeal as such. We may permit ourselves to speak of *space* provided only we take this expression to refer to the totality of all real physical bodies. 'But,' one may object, 'if a body can move, must there not be an empty place *into which* it can move?' Brentano counters: although we can say that the note C occupies a determinate place in the scale, it does not follow that there *is* a scale existing in its own right in which the note C is to be found.

* * *

We wish to express our gratitude to the following people and institutions: to Dr Reinhard Fabian of the University of Graz for his expert assistance in preparing Brentano's manuscripts for publication; to the late Dr George Katkov of St Antony's College in Oxford for the valuable advice that he gave to us during the preparation of the original edition; to the Felix Meiner Verlag in Hamburg for permission to translate the German edition of this book; and to the Franz Brentano Foundation which has supported the preparation of both the English and the German editions.

NOTES

. Kastil had begun this work during the time he was Professor at Innsbruck. He returned to the project, after a long interruption, in the summer of 1943. The works that appear here were prepared from the texts that Kastil had edited. Obvious omissions and typographical errors have been corrected. The manuscript titles and numbers that have been used here stem from the cataloguing of Brentano's *Nachlass* prepared in Innsbruck in 1951-1952. A brief description of this *Nachlass* may be found in 'The Manuscripts of Franz Brentano,' by J.C.M. Brentano, in *Revue Internationale de Philosophie*, Twentieth Year (1966), pp.477-482.

2. The term 'station' is used by Wicksteed and Cornford in their introduction to the English translation of Aristotle's *Physics* (Loeb Classical Library: Harvard and London, 1952).

3. 'The terms 'continuous,' 'contiguous,' and 'next-in-succession' have been defined as follows: things are 'continuous' if (while they are themselves distinct in the sense of occupying different places) their limits are one, 'contiguous' if their limits are together, 'next-in-succession' if they have nothing of the same nature as themselves between them. If these definitions are accepted, it follows that no continuum can be made up of indivisibles, as for instance a line out of points .... The points would have to be either *continuous* or *contiguous* if they were to make up a continuum.' *Physics*, Book VI, Chapter 1, 231a21 through 231b24; Loeb Library edition, Vol.II, p.93.

4. Richard Dedekind, *Stetigkeit und irrationale Zahlen* (Braunschweig, 1872). Compare George Cantor, *Gesammelte Abhandlungen*, edited by E. Zermelo (Berlin, 1932). There is no indication that Brentano was acquainted with Brouwer's intuitionistic conception of the continuum; but Brouwer's conception is much closer to that of Aristotle than to those of Cantor and Dedekind. The intuitionistic conception of the continuum is set forth in A. Heyting, *Intuitionism: An Introduction* (Amsterdam: The North Holland Publishing Company, 1971).

5. In the final essay, there are some critical remarks directed against Einstein's special theory of relativity. Brentano imputes to Einstein the thesis that time is the fourth dimension of space, but concedes that the thesis is a fiction which 'will in many respects prove harmless.' It is important to realise that, for the most part, the philosophical questions about space and time with which Brentano was concerned, are quite different from the questions of physics with which Einstein was concerned. The metaphysical theories of Brentano were not intended as contributions to physics.

6. Compare *Psychology from an Empirical Standpoint* (London: Routledge and Kegan Paul, 1974), p.354.

7. Compare *Psychology*, pp.355-6.

8. *Psychology*, p.157.

9. *The Theory of Categories* (The Hague: Martinus Nijhoff, 1981), pp.60-1.

10. When Brentano lectured on descriptive psychology at the University of Vienna between the years 1888 and 1891, he proposed the following account of the meaning of the word 'past.' The adjective 'past' should not be thought of as expressing a genuine attribute at all. Rather, it may express what Brentano called 'a modifying attribute [*ein modifizierendes Attribut*].' If we say of something that it is a 'tall King' or a 'wise King,' we imply that the thing *is* a King, and our adjectives express genuine attributes. The adjectives 'tall' and 'wise' in the expressions 'tall

King' and 'wise King' could be said to *add to* what is suggested by the noun 'King.' But if we say of something that it is an 'apparent King' or a 'supposed King' or a 'past King,' we do not imply that the thing *is* a King and our adjectives, therefore, are only 'modifying.'

11. This move will call to mind Kant's 'Copernican revolution' and the doctrine that time is a 'form of inner sense.' But Brentano's view can hardly be called Kantian. This latter point is obvious when we consider Brentano's doctrine according to which what is real coincides precisely with what is temporal or 'in time.'

12. In note 143 to Brentano's text, Kastil takes one passage to suggest that Brentano came to have doubts about this general view. But it is not clear to us that the passage needs to be so interpreted. In Edmund Husserl's *Zur Phänomenologie des inneren Zeitbewusstseins* (The Hague: Martinus Nijhoff, 1966), there is a criticism of Brentano's early views about our consciousness of time (cf. p.10-18). Husserl does not mention Brentano's subsequent view that our consciousness of time has its source in modes of presentation [*Vorstellungsmodi*]. See Oskar Kraus, 'Zur Phänomenognosie des Zeitbewusstseins,' *Archiv für die gesamte Psychologie*, Band 75 (1930), pp.1-22; reprinted as 'Toward a Phenomenognosy of Time Consciousness' in Linda McAlister, ed., *The Philosophy of Franz Brentano* (London: Duckworth, 1976), pp.224-239. Kraus's paper was occasioned by the publication of Martin Heidegger's edition of Husserl's 'Vorlesung zur Phänomenologie des inneren Zeitbewusstseins,' in the Husserl *Jahrbuch für Philosophie und phänomenologische Forschung*, Band IX (1928), pp.367-489. Brentano's lectures had been the impetus for Husserl's work on this topic, as Husserl notes (see p.xv of the 1966 edition).

A Note on Brentano's Terminology

Barry Smith

A number of obstacles are set in the way of understanding the text which follows, both by peculiarities of Brentano's terminology and also by his use of German constructions with no direct English equivalents. Thus a key role is played in Brentano's psychology by the notion of *Vorstellung*, here translated always and everywhere as *presentation* (with its derivatives: *to present, presenting,* etc.). As Brentano himself explains the matter in his *Psychology from an Empirical Standpoint*:

> We speak of a presentation whenever something appears to us. When we see something, a colour is presented; when we hear something, a sound; when we imagine something, a fantasy image. In view of the generality with which we use this term it can be said that it is impossible for conscious activity to refer in any way to something which is not presented. When I hear and understand a name, I have a presentation of what that name designates; and generally speaking the purpose of the name is to evoke presentations. (Vol. II, p. 34; trans. p. 198)

The notion corresponds in many ways to the British empiricist *idea*: to have a *Vorstellung* of something is, roughly, to have an idea of it, to have it before one's mind either intuitively or conceptually. *Idea*, however, has connotations which go beyond those of the more technical *Vorstellung*, and it lacks convenient verbal and adjectival forms. The use of *to present*, on the other hand, brings difficulties of its own, against which the reader is here forewarned. Thus we would normally say that it is the colour or sound that presents itself to us, whereas, on the technical usage here adopted, it is the perceiver who presents the colour or sound to himself. Moreover, there is a danger in a work on the philosophy of time that the verb *present* will be confused with *present* in the sense of *at the boundary of past and future* (as for example in a phrase like *making present*). This difficulty is here avoided, where necessary, by using *in the present* and its derivatives to convey the latter meaning.

The three classes of mental phenomena for Brentano are those of presentation, judgment and phenomena of love and hate. Difficulties are created in particular by Brentano's conception of judgment as a matter of the acceptance (*Anerkennung*) or rejection (*Verwerfung*) of an object given in underlying presentations. This is because *to accept* has connotations alien to those of *to judge as existing*, and where it seems desirable I have therefore used *accept* or *acknowledge* as a translation of *anerkennen* in order to emphasise Brentano's technical usage. *Erkennen*, on the other hand, is a term employed by Brentano here in a non-technical way and is translated variously as *grasp*, *apprehend*, and *see that* (as in: he sees that such and such is the case). Mental phenomena in general are classed together by Brentano as thoughts. Brentano's *Denken* is accordingly to be understood in the wide sense of Descartes' *penser*. A technical use is given by Brentano also to the term *Bemerken*, here translated as *to notice* or sometimes as *to perceive*.

Presentations, judgments and phenomena of love and hate have as their objects always and only things; indeed everything that exists, for Brentano, is characterised as a thing (in his somewhat technical sense). *Something*, in what follows, can therefore be read in every case as synonymous with *some thing*. German allows the formation of substantives simply by capitalisation of the appropriate adjective (*ein Rotes* from *rot, ein Ruhendes* from *ruhend*, etc.). For Brentano it must follow that in each case the resulting substantive, if it is admissible at all, is synonymous with the referring expression which results when the adjective in question prefixes *thing*. *Something red* (*ein Rotes*) would thus be synonymous with *some red thing*, and so on. Here, however, I have taken care to render the given substantives as far as possible by means of neutral forms such as *red something, something at rest, what is at rest, what is omnipotent*, etc., and have avoided importing *thing* (*thing at rest, omnipotent thing*, etc.) even where this would have yielded a more elegant translation. *Omnipotent entity* translates *allmächtiges Wesen*.

Since, according to Brentano, everything that exists is a thing, it follows that there are no facts or similar entities in the world as Brentano conceives it, and Brentano's German indeed avoids all constructions like *in virtue of the fact that, following on from the fact that*, which English can barely do without. Where such constructions occur in what follows they are therefore to be treated merely as turns of phrase, lacking all ontological significance. Since Brentano acknowledges no entities other than things, it follows also that he accepts only that mode of existence which is appropriate to things and does not distinguish between existence and subsistence in the manner of, for example, Meinong or Wittgenstein. *Exists* has therefore been employed in what follows equally as translation of *existiert* and *besteht*. *Continuing to exist* translates *Fortbestehen* and the substantive *Fortbestand* has been rendered occasionally as *perseverance*.

Things, for Brentano, are either *substances* or *accidents*. The latter expression sometimes causes trouble because of its ambiguity as between *akzidentell* (belonging to one or other accidental category) and *zufällig* (contingent, the result of mere accident). Things are divided also into species and genera, or in other words they fall under intersecting concepts which form trees of greater and lesser generality, from highest genera at the top to lowest species at the bottom. What distinguishes one species from another, within a single genus, is a specific difference, also in certain contexts called a *distance* or *interval* (*Abstand*). Some species are continuously differentiable, that is, their specific differences form a continuum (a characteristic which has nothing whatsoever to do with the differential calculus). To Brentano's scholastic terminology of species and difference there belongs also the terminology of *determination, peculiarity, mark* (*Bestimmtheit, Eigentümlichkeit, Merkmal*) and in particular of *local determination* (*örtliche Bestimmtheit*) or *determination of place*.

What is continuous (*das Kontinuierliche*) is divided, in Brentano's terminology, into what is continuously many, for example a spherical body extended in space, and what is merely continuously manifold, as for example the mid-point of a spherical body in rotation. What is continuous is divided further into *chronisch* and *topisch Kontinuierliches*, here translated as *temporally* and *spatially continuous*, respectively, though what is intended is a more subtle distinction between continuity of succession and continuity of juxtaposition. As Brentano shows, not everything that manifests continuity of juxtaposition is spatial in the commonly accepted sense.

Brentano's term *boundary* (*Grenze*) may cause problems because of the ambiguity in German as between *boundary* and *limit*. Since the first of these alternatives almost always yields the more natural translation, it is this which I have used throughout, even at the expense of occasional clumsiness. *Einer Grenze nach* I have rendered as *in a boundary* or sometimes *according to a boundary*.

Two great scholars of Brentano-inspired philosophy — Roderick Chisholm and Karl Schuhmann — provided valuable help in the preparation of this translation, and I am grateful to them both. Neither are responsible for errors which remain.

Part One

THE CONTINUUM

I. On what is continuous

Dictated 22 November 1914 *[Meg 24]*

This item was transcribed for publication by Kastil on two occasions (in 1943 and 1944). The texts of the two versions differ slightly, and the version that is reproduced here is based on Kastil's transcription from the year 1943. [Editors' note.]

1. The question concerning the concept of continuity cannot be framed in such a way that one would call into doubt whether we do in fact possess such a concept. For otherwise we would not be able to understand ourselves when arguing about other aspects of this concept. This implies also that we have attained this concept in the same way in which we are able to attain other concepts. Thus it cannot be inherited and given from the start, since this holds of none of our other concepts.[1] Yet we have been in possession of it long before we began to philosophise, and our memory can no longer tell us about the way it first developed. Fortunately however it is true of all our concepts that they are actually not given to us without interruption, so that the process by which they develop is in fact repeated even now. We can perceive this process and analyse it while it is still fresh in our memory.[2]

2. All our concepts are either taken immediately from an intuition or combined out of marks that are taken from intuition.[3] It has lately been asserted that it is in this second way that we arrive at the concept of what is continuous. One has, for example, pointed to the cases in which we insert fractions between two numbers, between 1 and 2, or indeed between 0 and 1. Thus for example we insert 1/2 between 0 and 1, and then continue halving by inserting 1/4 and 3/4, 1/8, 3/8, 5/8, 7/8, 1/16, 3/16 and so on. If one imagines this process of halving to be continued to infinity, then nowhere would there be manifested any gap of finite magnitude, and the

1

result would already be something that one could accept as an example of continuity. Yet this continuity would be incomplete, for there would still remain certain relations of magnitude which, even after a completely executed process of halving, would be represented by no inserted fraction, as for example that relation which is signified by the fraction 1/3. This then leads to the further requirement that we should proceed as with 2 so also with 3, 5, and all other natural numbers not yet implicitly used. We seem thereby to have achieved something that approximates much more to an example of something completely continuous. But still, the algebraic relations of irrationals would remain unconsidered, and thus in order to fill the still unoccupied positions one would have to carry out further insertions of the algebraic irrational numbers. Once this has been done, it is still of course not excluded that there would be more to be inserted, but then one would already be so close to the idea of continuity, of that which would leave no room for any sort of intercalation, that one could surely ignore this latter incompleteness. If, however, one did not want to do this, then one could construct the concept of the completely continuous by conceiving of something which — in contrast to all our examples of incomplete continuity — was such as to allow no further insertions. One would in fact need for this purpose simply to carry out a process wholly analogous to the one already considered. For even the halvings were not executed *in toto*; rather, the idea was formed of an execution which would differ from the finite execution actually achieved in being infinite. Thus one now forms, — in opposition to the groups of intercalations carried out *in infinitum* of all halvings, rational fractions, algebraic irrational fractions, transcendental fractions, etc., where there always remain positions open for new insertions — the conception of a complete realisation of this process, in relation to every possible relation of magnitude. There would then be given something which would present itself as a perfect example of continuity.

The attempted construction here described is similar to one that is to be found for example in the work of Poincaré,[4] though it is not identical therewith. Poincaré begins, certainly, with the insertion of all rational fractions between two whole numbers such as 0 and 1. He then goes on in the manner of Dedekind (see the Addendum on pp.39 ff. below) to the intercalation of all irrationals, calling the series so attained a continuous series of the second order. He then appeals to the fact that one can distinguish infinitely small magnitudes, magnitudes which are infinitely smaller than

2

other infinitely small magnitudes. (A point is related to a finite line as the line itself to a finite plane and this to a finite body, and in this way relations of magnitude can be conceived *in infinitum* with regard to ever infinitely greater magnitudes.) Thus, he believes, one can speak of a third, fourth, etc. order of continuity. This last idea seems to be peculiar to Poincaré.

It is also worthy of note that, although he credits some sort of validity to this construction, Poincaré nonetheless admits that it gives rise to a number of problems. Already, he suggests, the very possibility of rational fractions becoming infinitely smaller would presuppose that continua exist. And a further reservation seems to attach itself to the transition from the rational to the irrational numbers since one would be allowing oneself in the latter case to insert mere 'symbols'. Thus after dividing up the series of rational fractions which contain squares into those which are smaller than 1/2 and those which are larger than 1/2, one discovers in relation to the series of their roots, that the one part has no largest member, the other has no smallest. As a consequence of this, the two series of roots are supposed to border not on each other but rather on a cut, which is symbolised by $\sqrt{1/2}$. (In fact it seems unjustified to assert that a series without a last member can border on a series with a first member but not on a series with no first member. It is only through such an assertion however that Dedekind arrives at the insertion of the irrational fractions. If one takes the whole series of rational and irrational fractions, then one finds that no two follow each other immediately, so that the series differs in this respect not at all from the series made up exclusively of the rationals, and I do not see why it should be more unacceptable in the one case than in the other to say that one could divide a series in two parts of which the one has no initial, the other no final member. According to Dedekind the rational fractions would certainly admit of being dissected in this way, not however the irrationals.) Finally, Poincaré makes the striking remark that it can be established in no other way than through convention what magnitude should be ascribed to the distance between one fraction and another. In the case of a true continuum, however, the magnitude would be given independently of convention.

The idea of continua of various degrees of completeness seems also to be incompatible with the true solution to the problem of constructing the continuum. If one raises all the rational and irrational fractions between 0 and 1 to some power, then one obtains precisely the series with which one started but with a certain displacement, and the same holds where all members which are either rational or irrational are themselves raised to a certain power. In this way the magnitudes of the distances between the fractions appear not to be determined by the magnitudes of the fractions.

(Poincaré's two orders of continuity recall the two powers of Cantor. Yet the fact that there exist infinitely small magnitudes of a higher order cannot be regarded as a demonstration of a higher power, indeed the points of a surface for example are supposed to be of the same power as the points of a line, etc.)

(How, according to Cantor, is one to relate univocally the totality of the irrational points on a line to the totality of all its points, and how is one to relate the totality of transcendental irrational points to the totality of algebraic irrationals?)

3. Proceeding in this way, we should have to ascribe to the concept of continuity an origin in operations of thought both artificial and involved. This seems unacceptable from the very start, for how could this concept then be found in the possession of the simple man or even of the immature child? And further, how dubious it appears to suppose that the halvings and other divisions have been executed to an actual infinity, that they have been brought to completion, just because one can assume without absurdity that they have been executed beyond any arbitrarily determined limit. It is not to be denied that one is here accepting something simply impossible.

That one has indeed here posited something completely absurd is seen immediately if one splits the supposedly continuous series of all fractions between 0 and 1 into two parts at some arbitrary position. One of the two parts will then end with some fraction f, the second however could now start only if there were some fraction in the series which was the immediate neighbour of f, which is however not the case. With what, then, does the second series begin? With a multiplicity of fractions rather than just one? But this, too, is impossible since every fraction is distinguished from every other by a before or after in the series. But if not with a single fraction and not with a multiplicity of fractions then with what, since there is nothing to be found in the series other than fractions taken either singly or in groups? We should apparently have something that began but without having any beginning.

One sees that in this entire putative construction of the concept of what is continuous the goal has been entirely missed; for that which is above all else characteristic of a continuum, namely the

4

idea of a boundary in the strict sense (to which belongs the possibility of a coincidence of boundaries), will be sought after entirely in vain. Thus also the attempt to have the concept of what is continuous spring forth out of the combination of individual marks distilled from intuition is to be rejected as entirely mistaken, and this implies further that what is continuous must be given to us in individual intuition and must therefore have been abstracted therefrom.

Someone might however object at this point that the failure of the attempt here presented would not rule out the possibility that some alternative might have greater success, might even possess the advantage of greater simplicity. One could say that something continuous was present where a whole was given that could be thought of as divided not, certainly, into infinitely many parts, but still *in infinitum* into parts. The first-mentioned view would be absurd, and would be guilty of absurdity already in the idea of the completion of the totality of possible halvings; it would be possible to show further that with the completion of infinite halving the magnitude 1/3 would have to be arrived at, when, as has been correctly emphasised, this magnitude cannot ever be reached. Not quite so absurd, however, would be the idea of a magnitude which can be divided *in infinitum*. And nothing would be easier than to show how one arrives at this idea, since it forms the contradictory opposite of that which can only be divided into a finite number of not further divisible parts. One could easily see also how, having been taught by experience that in the division of bodies one never reaches any boundary that is not in all probability capable of being breeched, one is led by habit to take for granted the idea that every body is divisible. And since certainly every division of a body leads to parts which are themselves bodies, one is led further to believe in the possibility of a division *in infinitum*, even if not by any power of one's own. Of course this habitual assumption would as such not yet be justified from a scientific point of view. Yet science in no way contradicts it; rather it lends support to it in its own fashion by showing how the phenomena of experience can be reduced to quite simple general laws only on the basis of the hypothesis of the continuity of what is spatial. Without the assumption of continuity the assumption of a law of inertia would become untenable and the entire remarkable power of our mechanics would thereby be destroyed. Both the origin of the concept of that which is continuous and also its significance for the actual world would in this way be explained.

5

4. Even should one accept, however, that this theory avoids some of the cruder errors of the earlier view, it is still far from satisfying all requirements. Democritus believed in the extendedness of his atoms, not however in their physical divisibility, and it is a long familiar fact that when we distinguish parts in our thoughts we soon arrive at a certain limit. Moreover, everyone knows that in virtue of their small size corpuscles are far from being capable of being separately noticed, even if they are still very much capable of being divided physically. Yet we still award them continuity, and it seems therefore indubitable that being continuous and being divisible *in infinitum* are concepts that do not coincide in their content. We can remind ourselves also of what was said earlier about the peculiarity of boundaries and the possibility of their coincidence. He who does not show how we arrive at these ideas is not, either, allowed to flatter himself with having sufficiently clarified the idea of the continuous.

5. Thus I affirm once more, and with still less contestability, that the concept of the continuous is acquired not through combinations of marks taken from different intuitions and experiences, but through abstraction from unitary intuitions.

One does not need to search long, either, for the intuitions at issue, since, as I dare to assert and shall attempt to prove, it is much rather the case that every single one of our intuitions — both those of outer perception as also their accompaniments in inner perception, and therefore also those of memory — bring to appearance what is continuous. Thus in seeing we have as object something that is extended in length and breadth which at the same time shows itself clearly as allowing us to distinguish a front and rear side and thus as characterised as the two-dimensional boundary of something extended in three dimensions. And since this continuous something presents itself to us who see as being our primary object, we see also at the same time and as it were incidentally, our seeing itself, that is, we are conscious of ourselves as ones who see, and we find that to every part of the seen corporeal surface there corresponds a part of our seeing, so that we also, as seeing subjects, appear to ourselves as something continuously manifold. And still more, what appears to us first and foremost is rest and motion; so also persistence and gradual change appear to us as primary qualitative objects. This happens in that, whilst certainly in our perceptual presentation of the primary object we are never able to present the same place filled

6

with two qualities simultaneously,1* still we are able to present it as filled with one quality as present, with another as most recently past, and with yet another as further past, whereby the transition from present to further past takes place in an entirely continuous manner. Thus once more we appear to ourselves, in seeing phenomenal qualities following each other in a temporally continuous way or in seeing them persisting continuously in time, as something that is continuously manifold.

Given, therefore, that what is continuous is present in every intuition, the whole question as to the origin of the concept seems to have been dealt with in the simplest way. We have after all seen that this concept is gained not through any intricate process of combination but rather in immediate fashion through simple abstraction from our intuition.

6. Against this, however, objections will be raised from certain venerable quarters which we should not leave unconsidered. Thus it is said that according to general consensus it belongs to the nature of the continuous that it can be divided in thought over and over again *in infinitum*. Now it is certain, and we ourselves have conceded this earlier, that we are no longer able to distinguish with our senses particles which do not attain a certain size. It is therefore only in our thoughts that we are supposed to be able to undertake halvings and any other sorts of divisions. But who is to guarantee that in thinking this we do not go off entirely into the transcendent and do not lose ourselves in nothing but chimeras? Only in so far as we can distinguish with our senses does experience act as guarantor. Beyond that, we seem to be lacking in surety altogether. Our thought-apperceptions may not correspond phenomenally to anything at all, in which case the divisibility would not be a genuine divisibility but would be entirely imaginary.

Further, we have said earlier that the concept of a boundary and

---

1* In the intuition of a body I present the body primarily *in recto*. I can also however, by presenting to myself primarily a presenter of this body, present the body *in obliquo*. And it is in this way that I present it when I have a memory of an earlier experience which consisted in my presenting it in perception. It is not my intention to deny, by what has been said in the text, that I can in the same way, while presenting a given place *in recto* filled with blue, present this same place *in obliquo* filled with red. Also it seems as though I am able simultaneously to present the same place in obliquo filled in multiple fashion with colour-qualities. Similarly, when I judge something, I am also able to think more than *one* other judger judging about the same question in ways diverging variously from my own.[5]

the possibility of a coincidence of boundaries is essential to the concept of what is continuous. It was precisely the fact that nothing of such a boundary or coincidence of boundaries resulted from the attempts at combination discussed above which was to serve as proof of their total failure. But if this is so, then it must be demanded that the attempt to clarify the origin of the concept of the continuous via abstraction from what is given intuitively does not suffer from the same defect. And yet it seems obvious that it does, for if it is conceded that parts of a certain smallness are no longer noticeable, then this is to concede also that points and also other boundaries, all of which do after all dispense with all magnitude in certain directions — as for example a line should have no breadth — could least of all be capable of being distinguished in intuition. But then it seems also incapable of being established via intuition that boundaries and a coincidence of boundaries should exist at all in what is intuited.

7. These objections seem at first to be very plausible; yet the means to refute them are found if one inquires more closely into the peculiarities which, according to experience, our noticing has, and takes these into account. If we imagine a chess-board with alternative red and blue squares, then this is something in which the individual red and blue areas allow themselves to be distinguished from each other in juxtaposition, and something similar holds also if we imagine each of the squares divided into four smaller squares also alternating between these two colours. If, however, we were to continue with such divisions until we had exceeded the boundary of noticeability for the individual small squares which result, then it would no longer be possible to apprehend the individual red and blue areas in their respective positions. But would we then see nothing at all? Not in the least; rather we would see the whole chessboard as violet, i.e. apprehend it as something that participates simultaneously in red and in blue, though of course not, strictly speaking, in the same positions, since red and blue do after all, as contrasting colours, exclude each other mutually. Thus one would indeed be able to say that both red and blue positions were to be found therein. But one would not be able to go beyond this general determination so far as to be able to determine down to the last details whether this or that point would belong to the red or to the blue ones. We see, therefore, that the limitation of our capacity to differentiate what is indisputably involved here does not deprive us of the possibility of asserting with all certainty that the surfaces before us are here and here red, there and there blue.

Something precisely analogous, now, holds in our own case. Certainly we cannot distinguish the individual points and boundaries in the continuum that presents itself to us, just as we could not distinguish the individual red positions in the divided chessboard. Yet this does not hinder us in apprehending with complete certainty that boundaries and coincidences of boundaries are numberlessly present in the whole in question. The general character of that which is continuous, like, in the just-mentioned example, the character of that which participates in red and blue and is thus to be referred to as reddish and bluish, remains beyond all possible doubt. Differences in intensity in sensory phenomena of the sort that are revealed for example in the loud and the soft or in strong and weak smells, could also be called in aid to illustrate this law of noticeability. For if we investigate precisely how these differences in intensity are to be conceived, we find that we are dealing a certain 'density' of phenomena in the sensory field. Unnoticeably small parts of the whole may be filled or empty. This emptiness is then manifested in the whole as a kind of weakening or diminishing of the phenomenon. The fact that no particular gap is to be distinguished does not leave room for doubt as to the presence of gaps, in particular as in general.[6]

8. Having thus come to grasp the origin of the concept of the continuous we can now very easily make this concept clear and distinct. As with other concepts gained not through combination of various marks but through abstraction from a unitary intuition — as for example the concept of what is coloured — we have only to bring forward different intuitions which all contain the relevant mark and then perhaps, in order still more to draw attention to the crucial point, contrast these with others where this mark is either entirely absent or at least given only in a noticeably different way. I have reason to add this last caveat since of course, as I have already said, no intuition is entirely free of the concept of that which is continuous.

9. This now leads us naturally to the question as to the different respects in which something continuous would differ from other things and in terms of which a division into classes is to be undertaken.

10. a) One such respect, familiar to everyone, is that according to the number of dimensions. We distinguish one-dimensional, two-dimensional and three-dimensional continua, and recently it has come to be accepted that even the assumption of more than three-dimensional continua is in no way an impossibility. A continuum is

9

to be designated as one-dimensional if it has no other boundaries than such as are not themselves continuous. Thus for example a time is a continuum of one dimension, since its boundaries, as also the boundaries of its parts, themselves possess no temporal extension. We call them moments or instants of time. The spatial line, too, has no boundaries other than non-extended ones, namely the spatial points, and it is for this reason that Euclid defined the point as that which has no parts. The surface, in contrast, belongs with the two-dimensional continua since its boundaries comprehend not only points but also lines. And a body is to be designated as a three-dimensional continuum since not only is the whole body bounded by a surface but so also each one of its parts is separated from the remainder by a surface that is a two-dimensional boundary.

11. b) This leads us immediately to a second respect in terms of which what is continuous can be classified: we can distinguish continua which exist only as boundary of some other continuous thing from those which belong as boundary to no other continuous thing. The line and the surface supply familiar examples of the first. A time can serve as one example of the second, and a body, too, may be considered as belonging to this class, although we shall see later that there appear to be objections to this idea.[7]

If something continuous is a mere boundary then it can never exist except in connection with other boundaries and except as belonging to a continuum which possesses a larger number of dimensions. Indeed this must be said of all boundaries, including those which possess no dimensions at all such as spatial points and moments of time and movement: a cutting free from everything that is continuous is for them absolutely impossible. And this allows us to grasp very clearly the topsy-turvy character of the above-mentioned attempt at construction of the concept of the continuous through interpolation of fractional numbers, where every fraction is supposed to have existence without belonging to a series of fractions.

Boundaries, and among them also continuous boundaries, can be distinguished also however as inner and outer, as for example the mid-point of a solid sphere is an inner boundary, a point on its surface an outer boundary, and the same holds also of the top point of a cone. Thus also every surface which divides the sphere in two halves is an inner boundary, the surface of the sphere an outer boundary. This distinction is a consequence of the fact that for a boundary to exist it is required to belong to something continuous whose boundary it is, and to be connected with other boundaries of the same continuum. But it can nevertheless be the case that this connection should be missing on one or more sides. It suffices that it is not missing on the remaining sides. Thus life has a point at which it begins and a point at which it ends, and the former is connected only as earlier to the later moments, and the latter only as later to the earlier moments, while this sort of connection is given in both directions for the moments in between. One can say quite correctly that the inner moments of life exist both as final boundaries and as initial boundaries of the parts of life, but that the first moment of life exists not as final boundary but only as initial boundary of a part of life, while the last moment exists only as final boundary but not as initial boundary of a part of life. We refer to these differences as differences of greater and lesser *plerosis* of the boundary, and say of the beginning and ending points of life, both of which have a lesser plerosis than the intervening moments of life, that they are boundaries in relation to different and opposing sides.[8]

In other cases there is a still more manifold differentiation in relation to the sides and degree of plerosis of a punctual boundary. Imagine the mid-point of a blue circular surface. This appears as the boundary of numberless straight and crooked blue lines and of arbitrarily many blue sectors in which the circular area can be thought of as having been divided. If, however, the surface is made up of four quadrants, of which the first is white, the second blue, the third red, the fourth yellow, then we see the mid-point of the circle split apart in a certain way into a fourness of points. For in so far as it forms the apex of the blue quadrant it has one quarter of the plerosis that it had as blue point, since it was then mid-point of a completely blue circular surface and now coincides with one point belonging to a red, one belonging to a yellow, and one belonging to a white surface, of which each has the same degree of plerosis, though each in relation to different sides. Consider a boundary line of the blue quadrant, for example that which separates it from the white. This, too, has a lesser plerosis than a straight line separating one part of the blue quadrant from another. And if we speak of a line which forms the diameter of a solid sphere then we shall also say of it in the case where the sphere is made entirely of gold — as contrasted with the case where it is made of four parts of which one is of gold, one of silver, a third of copper and a fourth of tin, all of which touch each other in the given diameter and coincide there — that there holds of the plerosis in so far as it is the boundary of the golden part and is itself golden something wholly analogous to what holds in relation to the point of the blue surface according to whether it was an inner point of the blue circular surface or the external point at the apex of the blue quadrant.[9]

Because a boundary, even when itself continuous, can never exist except as belonging to something continuous of more dimensions (indeed receives its fully determinate and exactly specific character only through the manner of this belongingness), it is, considered for itself, nothing other than a universal, to which — as to other universals — more than one thing can correspond. And the geometer's proposition that only *one* straight line is conceivable between two points, is strictly speaking false if one conceives the matter in terms of lines of incomplete plerosis whose pleroses, even though they coincide with one another, relate to different sides.[10]

There is a question whether, when a solid ball rotates, the axis remains unmoved. The answer to this question is that the axis does turn, for it is first directed according to the plerosis directed to one side, later according to that directed to another side, until it finally returns to its initial position. The same applies also to the mid-point of a turning disc.[11]

12. c) This leads immediately to a third important respect in which to classify that which is continuous. We just now said that something continuous which serves as boundary could exist only as belonging to something continuous of a greater number of dimensions and only in connection with other boundaries of the latter. Boundaries require such belongingness and such a connection in order to exist at all. Someone could however object that there are boundaries which exist even though that continuous thing which they bound does not exist — so that one cannot speak of a belongingness or connection since nothing to which they would belong or be connected would have existence.

This objection touches on a peculiarity of that class of continuous things which can be called the temporally continuous, standing over against another class of what might be called the spatially continuous. An example of the first is provided by every continuous thing proceeding in time, of the second by every continuous thing existing simultaneously in all its parts, as for example a given line or surface or three-dimensional extension. Those things are temporally continuous which exist only according to one of their boundaries: it is this which triggers off the objection mentioned. Yet the objection overlooks something essential: if, in the case of some continuous process, only one of its boundaries exists, where it must be said of all the others that they do not exist, then the latter are to be denied only in this sense, not however in every sense. Some of them have a historical factuality; the others have a factuality that is removed from the present in the opposite direction. Even someone who says that something was a year ago is in a certain sense accepting or acknowledging this something, not in the mode of the present, certainly, yet still in the mode of the preterite or of the future, and these modes of acceptance manifest themselves in their totality as something continuous, to which the mode of accepting as present would then belong as one of its boundaries. Thus it is possible without contradiction to affirm of that which proceeds continuously that it exists only with regard to one of its boundaries and that this boundary is none the less such that it does not lack its connection with the remaining boundaries.

Of all that is temporally continuous it is time itself which above all stands in need of conceptual clarification. The most multifarious of conceptions have fought and still fight among themselves, and it may be that none of the earlier philosophers has succeeded in giving a fully adequate and exhaustive answer to the question as to the concept of time.

Ancient mythology went so far as to make of time, Chronos, a god, where others insisted that time as such exerts no effective influence at all. Many said that time passes; others affirmed in contrast that time itself exists in a changeless manner and that it is merely what is temporal that moves through it in the course of its existence. Among the more recent philosophers it seems that Kant and Schopenhauer stand opposed in this way. Some declared time to be something absolute, others that it was nothing but the totality of relations between things in regard to their 'before' and 'after'. Some wanted to perceive in time some one unified individual; others duplicated times according to the number of things proceeding in time, as for example Descartes when he identifies time with duration. What exists simultaneously, on this view, exists at the same time, not however strictly speaking at one time. Many wanted to conceive time as a completely uniformly proceeding change which would provide the measure for all other changes in relation to their before and after and to their changing rates, as also for the length of a period of rest. Aristotle held that he could designate the rotation that he ascribed to the firmament of the fixed stars as time in this sense, in so far as this rotation measures earlier and later; and this view, once carefully considered, does not stand in contradiction with his remark to the effect that if the soul did not

13

exist then there would be no time at all.[12] Leibniz, who no longer believed in this eternal, uniform rotation, still took account of the idea of time as a uniformly proceeding change which could serve as measure for all other changes, and since he did not believe that such a regular change was present among that which is actual, he therefore denied actual existence to time in this sense and allowed it only as something formed in our mind as an ideal standard. Augustine, too, saw in time a measure which we use as criterion for the succession of that which happens. This criterion is in the last instance to be found in memory, in which that which no longer is in actuality still continues to exist as thought (and even if it is thought of as a succession it is yet thought of simultaneously). Thus as something thought it has an extension which it does not in actuality possess, since of course past and future do not exist in actuality, and the present has no extension. In regard to the question whether time is a unified individual existing continuously, Augustine must be reckoned among those who answer in the negative. This is on the one hand because everyone has his own specific memory, and then also because he says that the century is made up of the lives which run their course within it. One could say that he is inclined with Descartes to identify time and duration, but with the rider that his duration has existence, strictly conceived, only in the memory and in the expectation of he who is aware of and comprehends what has been and is to come. That which distinguishes the earlier and the later, when they exist together as earlier and later, is referred to by Augustine not at all, however important precisely this point might be. Others however held the wholly unjustified view that it is a difference of intensity that provides the distinguishing mark. Yet others were of the opinion that just as there exist absolute local differences for separate points of the visual field, so there would exist absolute temporal differences for points which we behold as separate and thus also as succeeding each other within a field of temporal intuition. The proponents of this view split into two camps. In the one are those of the opinion that, just as our visual field always manifests the same local determinations, so also the field of temporal intuition always manifests the same temporal determinations. They say that this is why the present moment — which is one of the end boundaries starting from which our temporal field extends (strictly speaking, only in relation to the past), always appears as one and the same.[13] In the other, in contrast, are those who hold that the temporal differences, both for the present as also for the points of time removed therefrom, change continuously and that it is this which explains why, when earlier experiences appear to us in memory, some make themselves felt as belonging to the more recent past, others to the more remote, others to the very very distant past. The modification of the temporal differences of the present takes place, according to this view, always in one and the same sense, so that they grow continuously and gradually become extraordinarily significant, as though it were a displacement along a straight line. This second view is to be ascribed to Mach, while the first has been favoured by, among others, Marty. Both stand equally opposed to Aristotle's doctrine which apparently recognised no objectively given change in the case of simple continuance since he regarded local motion as the primary continuity existing in time and conceived such motion as copresent in every other continuous change. Others, too, for example Leibniz, think as he does that something could preserve itself unchangingly in *every* respect and indeed completely renew itself exactly as it was, so that the idea of continuous existence would involve no gradual change in the object at all. Suarez held in this regard that that which preserves itself unchangeably while other things are subject to change would be such that its duration would have no internal extent, but only an external one. One could therefore say only that during the period of the duration a part of a regular motion which is in some respect greater, in others smaller, had run its course (a doctrine which manifests one of the most astonishing scholastic distinctions.)

Yet the other views, too, contain equally unacceptable elements, which does not however prevent them each in a certain respect from pointing somehow to the correct theory. What has been said must however do justice to the source of these wide divergences of opinion, and allow them to expose preliminary doubts and objections.

13. If we ask what it is that collectively distinguishes the spatially from the temporally continuous, then as already stressed it is above all the fact that something spatially continuous, if it exists, exists as a whole. This applies to a line, a surface, a body. In these cases we have to do with a spatial juxtaposition of parts. Still, the realm of the spatially continuous extends further. It includes not only topoids of four or more dimensions[14] as these have been declared conceivable by the geometers, and their continuous boundaries (though these are not properly to be referred to as spatial), but also other cases of continuous juxtaposition. Thus for example when we look at a painting then in our seeing, too, we may distinguish parts

15

that correspond to the parts of the painting. And as this distinction of parts may go on *in infinitum*, so our seeing, too, manifests itself as something continuous that is also clearly such that it exists simultaneously in all its parts. But further, even when we present to ourselves a duration or temporal change, every part of our presenting corresponds to a part of what is presented; and even though we present these parts as existing in succession, this succession is still presented at one and the same time, and thus here, too, the presenting is made up of parts which exist simultaneously and forms something continuous which we have to count not among the temporal but among the spatial continua.

That which is temporally continuous, in contrast, is never such as to exist except according to a mere boundary, although after what has been said earlier we must insist that even here the boundary according to which it exists does not itself exist in isolation from its continuum. This is possible in virtue of the fact that the continuum is to be accepted as existing also according to all its other boundaries, but with a continuously varying mode of acceptance which distances itself further and further from the mode with which the existing boundary is to be accepted. This occurs on two opposing sides. The one side is that of the past, the other that of the future, and it is in the present that they come into contact.

It is generally accepted that a certain 'factuality' applies to that which is historical and to that which is to come. It is accepted also however that this factuality is to be distinguished from the factuality which we ascribe to what exists in the present. Some[15] wanted to ascribe to everything historical and to everything that is to come one and the same mode of factuality which, in reflection of the fact that they share in common that they do not exist, would appear as a lesser factuality than that of the presently existing. They have not considered the fact that in such a case the present would appear as a boundary existing without any connection to other boundaries and without belonging to a continuum. Though all curved lines share in common the fact that they are not straight, it does not follow that they are all curved in the same way. Indeed one may deviate from another just as much as this other itself deviates from a line that is straight; or the deviation may take place on the opposite side within the plane. And this applies also, clearly, to those modes of acceptance in which something is accepted but not accepted as being now present.

It remains however to investigate, in regard to this continuous difference in modes of acceptance, whether, like the opposition between affirmation and denial, it applies only to the judgment, or whether it is to be found already in the realm of presentation. In this case judgment, because it rests on a given presentation, would also participate in this difference, just as the differences which distinguish judgments in so far as they relate to different objects are also present already in the realm of presentation. And then it is not difficult to establish that already the realm of presentation does indeed manifest the difference of these temporal modes, for otherwise we would not be in a position to present to ourselves a movement. Confirmation for this is provided further by the fact that a difference of temporal modes manifests itself also in the realm of feeling — whose relations form the third fundamental class alongside those of presenting and judging — even in the case of those relations of feelings which involve no relation to a judgment. Thus one can wish that something should have happened, that it should now happen, that it will happen, without forming the least judgment as to whether the wish can or cannot be fulfilled.[16]

The just-mentioned trait constitutes for the temporally continuous a principal characteristic difference over against the spatially continuous, and it must be said that this is a difference not in the objects of presentation but rather in the modes with which they are presented, which contribute no less to the content of presention.[17] Now however it is also clear that what is temporally continuous often manifests to us a continuous variation in relation to the object. This is so, for example, in the case of local motion. The places which a moving point successively occupies differ from each other just as much as those which exist simultaneously on a line, being only in addition presented and accepted now with this, now with that temporal mode. Only where something simple is presented as continuing to exist does it seem as though this second sort of difference, which presents itself as a difference in the object, should be absent, and it is precisely this which seems to make up the contrast between remaining the same and changing. This second difference seems no longer to be required for presentation, since we have sufficiently shown through the reference to the difference of temporal modes how a temporal continuity can arise. And the fact that both the layman and the far greater part of philosophers do not believe in the difference seems to serve as support for the view that it is not actually involved in our presentations. To this can be added further that those who claim to find in presentation such a difference on the side of the objects even in the case of simple perseverance, are generally also those who have not recognised the difference we have pointed out between modes of presentation and acceptance for the temporally continuous. This is why they needed to accept some sort of variation, and this need seemed to them to be satisfiable only by a theory of a variation merely on the side of the object. The temptation to accept such a variation against all experience was then only too great. There speaks against such a theory also the fact that, as we saw, its advocates were not able to agree among themselves as to how to specify the differences of objects which they propound. Their divergence extends so far, indeed, that according to the one the difference which marks out the object present from the object past should be the same for all times, where for the other the peculiarity of the present object is as such continuously changing and would in the course of time manifest the most considerable differences. And when one looks more closely, then one sees that both assumptions lead equally either to contradictions with experience or to still greater inconsistencies. According to those who believe in a differential change in what is present it would have to be the case that, when someone wakes up, the difference between the past and the present must make itself felt to him as rather considerable in virtue of the temporal distance involved. Thus there could obtain no total uncertainty about the length of sleep, and still less would it be possible that in cases of so-called split consciousness there should be a return of a so-called normal period with what appears to be an uninterrupted connection with a moment of time lying far away. The alternative view, however, which conceives what is temporally peculiar to all present objects as being such as to appear the same for all time, must admit at once that it does not always correspond to the temporal peculiarity which in fact pertains to existing things and that it is in all probability distinguished therefrom by an enormous distance. It could be accepted of this distance however only with great artificiality, and precisely because it can be of arbitrary magnitude, that it is the same in relation to all our senses. Yet there is not the slightest difference between the senses which makes itself noticed in this regard. One gets into further difficulties if one takes into account in addition to outer perception, which is lacking in evidence, also inner perception, which admits of no illusion. If, in seeing, we also perceive ourselves as seeing, and indeed as seeing now, then there is no way in which that temporal difference of the objects which is peculiar to what now exists could be eclipsed by another one not directly related to it lying further back. Yet

18

differences are manifested here as between inner and outer perception just as little as among the senses when compared with each other. The supporters of this view have therefore been forced to make further, still more extravagant assertions, which in fact contradict all experience. Indeed they go so far as to deny entirely our accompanying inner perception its character of being a perception of something as present, or indeed they reject it altogether. This necessity to reject something that stands secure above all other things is sufficient proof of how desperate is the situation of the proponents of the given theory. Thus we can regard it as established that a temporal difference is never manifested as a difference on the side of the objects.[18]

14. One should not, however, misunderstand what has been said here. If I regard it as established that, wherever we present something as continuing to exist, no object-differences appear in our presentation, but rather the content varies continuously through continuous variation of the temporal mode of our presentation, then this is not to say that the actual continued existence of the thing could occur without any variation in its reality. Not every real difference must appear in presentation, for the latter is often, if not always, universal. Thus for example we never have any knowledge of ourselves in regard to what distinguishes us as individuals from someone else. All our presentations, judgments and emotions as we perceive them in ourselves could just as well be had by another who would perceive them in himself. There must be some substantial difference between him and me, but there is apparent to neither of us even his own individualising substantial difference. Thus, to return to our question, it might very well be that, attached to all those things which exist in some one moment of time, there was some real specific difference which distinguished them from all things that had existed earlier or would exist later, but one which remained entirely transcendent to us. Indeed we can affirm with confidence that this is not merely possibly but actually and necessarily so.[19]

I want to demonstrate that this is the case above all for that being in relation to which it has normally been held to be excluded completely, namely for the first, immediately necessary principle of all things, which was recognised already in antiquity as a divine, all-knowing and all-determining being. As omniscient it must know all truth; it can of course however not know more than all truth, and so if a change should occur in the domain of what is true, then there must also occur a change in the divine intellect. But this is the case whenever anything changes. For then something is which earlier was not, and he who knows everything must now know that it is, where earlier he must have known in virtue of his omniscience not that it is, but that it will be. Consider carefully what I am saying: it would not suffice for God's omniscience for him to know the whole succession of worldly events without knowing what point in this development had been reached and which events already belong to the past while others still await realisation in the future. Thus we have to assume of the divine intellect that he presents and accepts everything that happens, from all eternity, but with a constant and uniform infinitesimal change in mode of presentation and acceptance. And we must say the same of him also as a willing being. His preferences have, to be sure, been completely decided from eternity to eternity; but what he once had willed as lying in the future will later be willed as present and then still later be appreciated as something that was. And in relation to this change, too, we have to do with something that runs its course in a continuous and uniform fashion. Yet in relation to God there is no composition of substance and accidents. Thus his every continuous change in thinking and willing is to be conceived as a substantial change. Hence my assertion is fully confirmed first of all in relation to God, that the things, however much they remain the same, are yet subject to a constant temporal real change in virtue of which they never exist in a later instant just as they were in an earlier.

But now the same conclusion is established also for all other things to the extent that these can exist only in virtue of the fact that they are sustained in existence by the divine principle. For clearly, in every new instant of their existence they will be sustained by God according to the corresponding simultaneous moment of the divine life, and this must stamp them with a peculiar character-trait which must be common to all simultaneously existing creatures. Thus we have a real temporal change also for all created things which corresponds exactly to the process of succession within God. All things change in this respect just as infinitesimally and completely uniformly as God himself, even though in other respects they may manifest very irregular variations or even at times allow no change to be detected at all, as when a standstill interrupts a local motion. This type of continuity, too, or rather these continuous things, belong to the class of the temporally continuous and are to be referred to as the temporal peculiarity — or also as the peculiar duration — of every thing.[20]

Already Trendelenburg, in contrast to Aristotle and other great thinkers of earlier times, had recognised that a completely changeless first principle could not bring about a change in that which it conditions. This is what he is getting at when he says that motion could perhaps lead to rest, but never rest to motion. In employing a formulation of this sort, however, he did not take account of the fact that rest is itself inconceivable without some change or other. We speak of a longer or shorter rest and of rest which begins or ends earlier or later. Its being longer or shorter demands, however, a plurality of parts and, according to the principle of indiscernibles, no multiplicity can exist without differentiating marks. For an earlier or later beginning, however, the *ratio sufficiens* would be entirely lacking without the assumption of a difference in the cause. What we just established regarding the first, divine principle resolves the present puzzle, and it seems that this theistic idea offers the only possibility of resolving it.

So much may here suffice to distinguish time and the temporally continuous from what is spatially continuous.

15. d) We come now to a very important respect in which to classify continuity. It is that which follows from the multiplicity of most if not all of what is continuous. Imagine, for example, a coloured surface. Its colour is something from which the geometer abstracts. For him there comes into consideration only the constantly changing manifold of spatial differences. But the colour, too, appears to be extended with the spatial surface, whether it manifests no specific colour-differences of its own — as in the case of a red colour which fills out a surface uniformly — or whether it varies in its colouring — perhaps in the manner of a rectangle which begins on one side with red and ends on the other side with blue, progressing uniformly through all colour-differences from violet to pure blue in between. In both cases we have to do with a multiple continuum, and it is the spatial continuum which appears thereby as primary, the colour-continuum as secondary.[21] A similar double continuum can also be established in the case of a motion from place to place or of a rest, in which case it is a temporal continuum as such that is primary, the temporally constant or varying place that is the secondary continuum. Even when one considers a boundary of a mathematical body as such, for example a straight or curved line, a double continuity can be distinguished. The one presents itself in the totality of the differences of place that are given in the line, which always grows uniformly, whether in the case of straight, bent or curved lines, and is that which determines the length of the line.

The other resides in the direction of the line, and is either constant or alternating, and may vary continuously or now more strongly, now less. It is constant in the case of the straight line, changing in the case of the broken line, and continuously varying in every line that is more or less curved. The direction-continuum here is to be compared with the colour-continuum discussed earlier and with the continuum of place in the case of rest or motion of a corporeal point in time. In the double continuum that presents itself to us in the line it is this continuum of directions that is to be referred to as the secondary, the manifold of differences of place as such as the primary continuum.[22]

16. Between the primarily and the secondarily continuous there obtain significant differences which we must not lose sight of. I emphasise in particular the fact that, in the case of primary continua, a certain uniformity is to be encountered throughout and as a matter of necessity, where it is present in the secondary continua only as a matter of exception. If in a case of motion the variation of place is now rapid, now moving forward almost unnoticeably and now interrupted by a complete rest (which still, as the extreme of deceleration, will be counted by the physicists as a case of motion), then the *temporal* change which underlies motion and rest shows itself as being without any increase or slowing down in its rate of variation. This has long been quite generally recognised, and it is only in our own day that people have allowed themselves to be confused by Einstein in regard to this evident truth. According to Einstein's view, there would, between the same two temporal points and reaching from the one to the other, be times differing quite considerably in length. Our healthy common sense, which is here evidence itself, raises its head in violent protest. But the controversy still continues to rage, which would surely not be the case if one had paid more consideration to the peculiar character of the primary as contrasted with mere secondary continua. If there could be times of different lengths between two points in time, then also there could lie between two points in space two straight lines of different length, and the geometer's method for establishing the congruence of triangles through coincidence of surfaces laid one upon the other would be rendered entirely invalid. It is not invalid however, because the continuous manifold of points that is presented to us in spatial lines and surfaces, like the manifold of moments of a stretch of time, belongs to the primary continua.

Still more widespread than Einstein's error is the idea that one could not speak at all of velocity in relation to time, since this concept would have application only in relation to changes of other sorts. A velocity, as it proceeds in time, is called quicker or slower according to whether the time it takes is shorter or longer. In order to have a velocity, therefore, a time should not be a time, but *in* a time. However, by 'velocity' we are to understand in the end nothing other than the rate of variation, and certainly it cannot be denied that wherever variation exists, it must exist in some determinate degree or other, whether this be constant (constant necessarily and without exception), or now higher, now lower. In the case of time, now, there exists a variation. If something red exists for one hour, then as something red it has suffered no specific alteration. But as something temporal it differs from what it was earlier by a whole hour. Therefore also it is indubitable that there is some rate of temporal variation. What should be denied is only that, like the degree of other variations, it could come to be lowered or raised. And something quite similar holds also of the spatial as such. For it is undeniable that this, too, varies, and a red surface the redness of which shows not the slightest deviation into the furthest distance nevertheless shows itself to be quite considerably different in its spatial determinations when I compare its beginning and its end. And just as it is certain that there exists here a variation, so also it is certain that this must have a determinate rate of variation, and one should not let oneself be swayed from recognising it because it appears throughout as completely the same, i.e. because it is incapable of any increase or decrease. Now space, in contrast to time, manifests a plurality of dimensions. It allows us to distinguish boundaries which are themselves continua and which proceed in multifariously different directions, now straight, now curved. Because of this we must here expressly add that the the degree of variation in every direction, no matter whether we are concerned with straight or with curved boundaries, is and remains always and without exception the same. This is why, when straight lines are the same in length, there is the same distance between their beginning and end points. In the case of curved lines this is not the case, though if I imagine such a line as having been partitioned into ever smaller and smaller parts, then the magnitude of the line approximates into infinity the magnitude of the sum of the straight lines drawn as arcs between the points of the partition, and this proves that the degree of variation in the case of curved lines must be exactly the same as in the case of straight lines. Not even the smallest difference could be specified which would not reveal itself as excessively large.

How different things stand in the case of the secondary continua! We referred just now by way of comparison to the often significant differences among the degrees of variation of lines and surfaces and also among the velocities of a motion. The differences in degree of variation which can be manifested by a coloured surface may also serve as intuitive examples for someone who actually believes in continuous variation here (as is commonly the case). Consider a rectangle which begins as pure red on the one side and, after proceeding through a continuous series of nuances of violet, ends up as pure blue. This transition can be here quicker, there slower, and even if it were completely regular from right to left the surface would still show a strong difference in regard to its degree of variation in this direction, while in regard to that from top to bottom this degree of variation would reach the null-point. However, every line conceived as proceeding in a skew direction, for example diagonally, would already show a lesser degree of variation. And in the case of curved lines the degree of variation would change irregularly.

17. In addition to the temporal as such we have also put forward the spatial as such as an example of a primary continuum. We have however, in speaking of motion and rest, described variation of place as being itself secondary in relation to temporal variation. One might see a contradiction here. Note however that we had in the first place spoken of space as conceived by the geometer who abstracts from its perseverance in time and therefore also ascribes to bodies three dimensions. If, however, he took into account the time in which it exists, then he should have to call it four-dimensional, as was noticed already by Lagrange and has been emphasised also by later thinkers. If we do not carry out this abstraction however, then we shall still have to go on affirming that the body does not appear in the fourth dimension thus accruing to it as primarily continuous, as it does in the remaining three dimensions. Rather, it appears as secondarily continuous, in that time running its course provides the primary continuum of a body which extends from the beginning to the end of time as at rest or as more or less in motion. With regard to this, its temporal dimension, it manifests all characterising marks of the secondary continuum, while it is just as clearly the characterising marks of the primarily continuous which appear in it in relation to the three remaining dimensions. These it possesses already in every single moment and

24

they therefore require no temporal extension to underlie them. This might at first glance cause discomfort. It is however just as consistently possible as is a property which, regarded as something that is given in a single instant, manifests no extension at all, yet regarded in its temporal perseverance appears extended and indeed as secondarily continuous.[23]

That which is spatial manifests differences in degree of variation. It has a variable teleosis (velocity of change), only in relation to its fourth dimension, as a boundary of what is three-dimensional, not however in relation to the way things stand within those three remaining dimensions in which the body presently exists. This is similar to the way in which one can speak of a complete teleosis in the vertical dimension in the case of the colour-rectangle varying in the horizontal direction while remaining constant in the vertical, even though each vertical line manifests an incomplete teleosis in every point in so far as it is a boundary in other directions. Hence a body does in fact have the character of a primary continuum in the three dimensions in which it presently exists.

Yet still one would have to admit that if our determinations are to be of complete exactness then the temporal continuum is in the eminent sense the primary continuum if compared to the spatial. This is not only because of the limitation of the primary character of what is spatial to three of its four dimensions, which, in so far as it is primarily continuous, allow it to appear in fact only as a three-dimensional boundary of something four-dimensional. It is also because of a certain multifariously different degree of variation which, precisely because what is spatial exists in these four dimensions only as boundary, can apply to it even in an individual moment of time. Moreover, it exists in an individual moment in not quite the same way according to whether it exists as boundary of something that is continuing to exist or of something that is gradually passing away, and, considered from the side of the past, whether it exists as something that has existed just as it is until now or has come to be what it is through gradual change. These are differences of temporal teleosis, which have significance also for the individual boundaries of time.

18. It is necessary that we become clear just as in regard to the concept of plerosis, so also in regard to the concept of teleosis, a concept which is of the utmost importance for the understanding of the continuum and in particular for the phenomena which occur in the case of multiple continua. For the moment we shall seek only to indicate what is involved in this matter by raising the following

question: Suppose a coloured disc varies regularly from one radius to the other, in that, after proceeding through one semicircle, it has passed from pure red to pure blue and then again from pure blue back to pure red. Would then every radius manifest from the centre to the periphery completely the same colour-nuance, or one that somehow varied? One seems commonly to be of the opinion that for every radius the colour-nuance is the same from the centre to the periphery; and therefore one calls, e.g., one of the radii pure red and another pure blue. However, a more careful investigation shows that this opinion is marked by contradictions and that it does not take account of the fact that every boundary, because it exists only in the context of the continuum to which it belongs as boundary, must itself show up differences in reflection of differences in this continuum. Hence red, as beginning of a slower or faster variation towards blue, cannot be red in the same perfection; and still less can it be as perfectly red as in the case where it belongs as boundary to a uniformly red surface. And thus also too, a body is not at a place with the same perfection or completeness when it is passing through it slowly or quickly as when it is at rest.

The circumferences of two circles stand to each other in the same ratio as do their respective radii. Hence it would be impossible for the totality of colour-nuances manifested by the disc here described on the circumference of that circle which lies at a distance of one half radius from its centre to fill more than half of the outer circumference of the disc. Of the colour-nuances of the outer circumference, on the other hand, one could manage to lodge only one half within the concentric circle half-way to the centre. Quantitatively, therefore, they stand to each other in the ratio 2:1. Yet clearly, if the individual nuances were the *same*, then since every nuance of the one corresponds to a nuance of the other, they would have to be of equal number. Still, however certain it is that they are dissimilar, they are at most infinitesimally so. Otherwise it could not be that the infinite multiplication that is given across a whole circumference should yield only a finite ratio for the circumferences of the two concentric circles.[24]

If we ask whether a line with changing teleosis, such as is manifested by that line which is the red radius of our disc, is equal in length to a line beginning and ending at the same points but belonging as boundary to a purely red surface, then we shall have to answer in the positive. Hence also a body that fills space is to be called equal in volume whether it is at rest or moving with an

26

arbitrary velocity, and this is enough to justify the geometer's abstraction from the differing teleoses of a body which remains constant or increases or decreases in size. And thus we also believe that we lose nothing if we allow space to count as a primary continuum and disregard that which, when examined closely, it lacks in this respect. (Indeed we shall perhaps not shrink from saying of the rectangle that begins with a red line and ends in a blue line at the opposite side after having continuously changed its colour that in the perpendicular direction it shows itself to be in full teleosis in respect of its colour. And equally, we shall not shrink from saying of our coloured disc that every colour-nuance from the circumference to the centre shows itself in full teleosis, in that, whatever change in teleosis there is for a given radius in virtue of its belonging to the disc, this applies to it only because it belongs to the surface in relation to the direction of breadth.)[25]

In order more fully to appreciate the distinguishing peculiarity of the above-mentioned fourth (temporal) dimension of the body, consider also the fact that, in the case of a body moving along a straight line, this dimension corresponds in no way to the distance of the place to which the movement leads from the place from which it began. Rather it corresponds always to the length of the time within which the motion takes place. In the case of very rapid motion the spatial distance between end point and starting point will be relatively large; when the motion is slow it will be small; and when the body is at rest it will in fact be nil, even though the body has in each case precisely the same extension in its fourth dimension.

Because the movement of a body can be conceived as constantly varying in velocity and both as slowing down to the point of total standstill and as increasing and multiplying *in infinitum*, so it is clear that in this fourth dimension of the body every degree of teleosis is conceivable and thereby also that degree which applies to the body in all its other dimensions. Then, however, for a motion in a straight line, given the same spatial distance between starting point and end point, the extension of the body in its fourth dimension must correspond to the spatial distance between the boundaries of the motion. Thus it appears that this spatial distance would be equal to the temporal distance between the beginning and the end of the motion. Thus it is indubitable that one can speak of a ratio of lengths between a time and a line, which leads to the recognition that, just as we can speak of concrete or designated and abstract or undesignated numbers, so also we can speak of concrete and abstract continuous magnitudes. Just as one can say that the number of three trees is that number which the believer ascribes to the three persons of the Godhead, so one can say in all truth that a certain time has a length of precisely one meter, whereby one is of course not affirming that we are in a position to use the meter to measure the time in question.[26]

19. We have drawn attention to a certain uniformity as characteristic for the primary continuum, a uniformity in degree of teleosis that it always and necessarily has. This is never the degree of a complete teleosis, but it differs therefrom always and necessarily to the same extent. We have also however to draw attention to another uniformity manifested by everything primarily continuous that is not merely a boundary. This consists in the fact that all primary continua are straight or, as one also says, plane or flat continua. Thus time as such proceeds in such a way that the distance from starting point to end point grows, as in the case of a straight line, in precise correspondence to the ratios of the lengths of time. Everyone is immediately inclined to agree with this assertion, and here, too, it was only in our own time that it was fated to be called into question by those who argued that in certain other cases, too, a slight curvature that is factually present remains unnoticed. However this is something that can lead to uncertainty only in the case of boundaries. Where we have to do with the interior of a continuum, every point has full plerosis, i.e. is connected in every conceivable direction with the relevant continuum. Now, since time proceeds in only one dimension, every temporal point within the interior of a time has a two-sided plerosis only, and thus other temporal points can be at a distance from it in only two directions. Yet this would not be the case if it proceeded with a curvature, since in the case of every curved one-dimensional magnitude there are to be distinguished infinitely many points each of which goes off from the starting point in a different direction. If they did not do this, then every intermediate point would lie exactly between starting point and end point, and that is the characterising mark of the straight line. Thus a time (which certainly is not a mere boundary) every internal point of which possesses full plerosis, can be nothing other than a perfectly straight continuum. It can be shown in practically the same way that space is a plane continuum in this sense, in that, like time, it does not belong to a four-dimensional continuum merely as a boundary. Now of course we have seen that this is strictly speaking the case in so far as, in regard to its persistence in time, it acquires a fourth dimension in

relation to which it appears as a secondary continuum. However it is clear that it belongs in regard to this fourth dimension to the temporal continua which are marked by the fact that they exist only in a single boundary — that boundary which falls within the present. Since however this boundary is one and the same for all the points of the geometrical body, no irregularity can enter in as a result of the fact that the body as boundary exists in connection with that which exists earlier or later. Hence just as it must be perfectly regular in regard to its teleosis, so it must be perfectly straight (a flat continuum) in regard to the ordering of its parts, just as though it existed in isolation from all connection with the earlier or later and was already in its three dimensions possessed of full plerosis. For this reason, too, one pays no regard to the extension of space through time, in relation to the now so much discussed question whether space is flat or curved.

To rule out all misunderstanding, I wish to point out explicitly that when I assert that space could not be other than flat, I am far from denying the possibility that there could exist topoids of other numbers of dimensions in addition to three-dimensional bodies. If there did exist a topoid of four or more dimensions then there would be found in and on it three-dimensional boundaries which could just as well be curved as straight continua. But they would just as little be bodies in virtue of their three dimensions as a line is a time in virtue of its one dimension. For they would, after all, exist only as boundaries of something not temporally but spatially continuous, and this cannot be assumed of a body without contradiction.

20. It is easy to understand that in the case of what is secondarily continuous one can speak of the just-discussed uniformity just as little as of a uniformity of teleosis, for in the case of complete teleosis it would have a specific variation in not even one of the conceivable directions. One thing can be said of it as of the primarily continuous, however, namely that, whether it is given in complete or in incomplete teleosis, it can never be limited in its extension in such a way that it manifests itself as an isolated boundary. Thus for example in the case of a solid sphere at rest it is impossible that the surface alone or one of the circles distinguishable on this surface should rotate with a velocity of one meter per second. It cannot happen either that a body should be at rest for one instant which would have been moving up until this instant with a velocity of one meter per second and would then instantaneously move off with the same velocity in the opposite

direction. What we must say in this latter case is rather that the two movements in opposite directions touch each other in time and that in one and the same present moment, whose plerosis is two-sided, the body experiences two opposite motions in half plerosis, each having a plerosis in relation to a different side. This contains no contradiction; indeed, when determining the notion 'at the same time' in the formulation of the law of non-contradiction one must add to the other clauses the condition that the plerosis of this 'at the same time' should be on the same side.[27] Aristotle and then the scholastics did not recognise this, which led them to other involved but still inadequate and quite arbitrary theses, in particular to that according to which there exists no first and no last point of a motion. Rest however was not supposed to lack such points, and this led them to the absurd conclusion that a body thrown up into the air and then falling down again should remain in total rest, between the time of ascent and the time of descent, for some time that is for us unnoticeably small. It was affirmed that when something that is ceases to be, then there was a last moment in which it was, but no first moment in which it was not. With a better understanding of the peculiarity of partial plerosis one should have said that in the same moment it exists in partial plerosis and that it does not exist in full plerosis,[28] since until that moment it had existed and from that moment on however it existed no longer. Thus the given temporal point applies to it only as its external, not as its internal boundary; and this will have to hold now quite similarly in relation to motion and rest.

21. Before concluding our consideration of the primary and secondary continua and of the distinctions of teleosis, we want to consider briefly a question that is, when properly answered, able to throw further light on these matters. Imagine a sphere which is at one moment fully at rest in a given place, at a later moment rotating within the boundaries of this place. In the first case it is clear that in regard to its temporal dimension the sphere exists in full local teleosis. What, however, is to be said of the other case? Clearly we have to give the same answer as we would give if only one of the plane circles that is divided into two halves by the axis of rotation were to be caught up in a rotation of exactly the same speed as is the whole sphere. The local teleosis in the temporal dimension appears therefore to be incomplete in the case of a rotating sphere, and this incompleteness increases with the velocity of rotation. The rotating sphere in no way comes off worse than the resting sphere in respect of its spatial reality; but one cannot deny

that the local determination (determination of place) of each of its parts, and therefore also of the whole sphere, is constantly different. The local determination of each individual point is changed not as to the degree of reality but as to the species of reality. The change is however of such a kind that, to the extent that it becomes more dissimilar to the local determination of the given point at the time of rest, the local determination of other points becomes more similar to it.2* And thus, if one attends to the whole of the sphere, then one is presented with something that seems, as far as local determination is concerned, to be completely equal to the sphere at rest, but is perhaps better not referred to as the *same* but rather as *equivalent*. One sees that the teleosis of something has a relation to its genesis and to the cause through which it comes to be. As to teleosis, continual perseverance and continual renewal in continual decay do not yield the same result.

22. We have pointed to the unacceptability of the scholastic assertion according to which a body thrown upwards must remain at rest through a certain small period of time before beginning its descent. It is however correct to say that the body is at rest for an instant. Such an instant of rest can occur perfectly well in conjunction with an infinitesimally small beginning and increasing velocity and also with an infinitesimally small ending and decreasing velocity of motion. The moment of rest will be a moment of rest in the same sense in which a red line in a coloured surface which varies in regular fashion from pure blue to pure red and then back to pure blue is truly to be called pure red. It remains however no less correct that it is still to be distinguished from a pure red which belongs as internal boundary to a purely red surface. Indeed, certain differences will still exist in the manner in which the momentarily resting body is at rest, according to whether the motion which leads thereto and departs therefrom is subject to a more or less strong acceleration. One is reminded in this connection of what was said about the purely red radius of a disc which proceeds regularly from the purely blue to the purely red radius and then back again to the radius that is purely blue. This

2* For the interpretation of this passage see A. Kastil, *Die Philosophie Franz Brentanos* (ed. F. Mayer-Hillebrand, Bern: A. Francke, 1951), p.156. 'The determination of place of every point is changed, not according to the degree of its reality, but in its species, and the change is such that, to the extent that it becomes more similar to the determination of place of the given corporeal point at rest, the determinations of place of other points become more dissimilar to it.' [Editors' note.]

radius is certainly purely red; but still, it has its red colour in greater completeness in the circumference than in the middle of the radius and suffers also a constant decrease from there to the centre of the circle. This is a decrease in completeness of its teleosis. And now similarly, as we see, one can speak also in the case of what is truly at rest of a greater or lesser completeness of its teleosis. When it rests momentarily it is truly at rest, but at rest in a more incomplete teleosis than when it remains at rest for a period of time. And the completeness of teleosis in what is momentarily at rest decreases also if the motion to which it leads in infinitesimal fashion accelerates more powerfully from the very start. If the earth were to attract with a doubled strength, then the moment of rest between ascent and descent would possess only half so large a teleosis. If the backward motion, for example in a case of impact, should occur immediately with finite velocity, then the momentarily achieved rest would exist not merely in incomplete teleosis, but also in incomplete plerosis. The second half of the latter would be missing and there would be given instead a motion in half plerosis.

23. e) What is continuous can be sub-divided in yet another important respect into what is continuously *many* and what is continuously *manifold*. We want first of all to make this important difference clear by means of examples. As an example of the continuously many we can give a body, of the continuously manifold someone who sees something spatial precisely in so far as he sees it. The body is a unity which can be decomposed into a plurality in such a way that if one of its parts is destroyed the remainder can continue to exist just as it was before. For this reason one could already now call it a duality as well as a unity — indeed one could say that it is a hundred, thousand, or arbitrarily large number of bodies, though it seems that one could not say that it was infinitely many, since it is certain that every body must have an extension and an extension of determinate finite magnitude. Every one of the parts distinguished in such a decomposition has nothing in common with the others; it is thus adjoined to the others in true summation as something totally new. Things are quite different in the case of someone who intuits something spatially extended. This someone is *as such* not something simple but something manifold, since he sees not merely one but many parts of a continuum and could go on to see one such part while he ceases to see the others. But in so far as he sees the one part he does not amount to something totally other than what he is in so far as he sees the other part. We have before us not a duality, as we would

have in the case where it was one who sees this part and another who sees that. It is not the case that in the union of the two the one who sees the one part is adjoined to the one who sees the other part as something totally new and other, as it must be in the case of a summation of unities. If a and b are two and b and c are also two, then the union of a and b and c yields not a fourness, as would be the case in regard to the summation of one two with a totally other two. And thus also in the case of the union of the one who sees the one part of what is spatially continuous with the one who sees the other part: this does not yield a true duality of intuiting subjects. Now someone might hold that one has nevertheless to do here not merely with something continuously manifold but with something continuously many, as soon, that is, as one speaks not *in concreto* of the one who intuits something spatial but rather *in abstracto* of the intuition of something spatially continuous, which is then capable of being decomposed into completely different units. This is however to overlook the fact that the detaching of the intuition *in abstracto* from the one who intuits *in concreto* is in every respect impossible; it can be carried out neither in reality nor even in thought. Such detachment can occur only one-sidedly, in the sense that the person of the one who sees can be conceived as continuing to exist even after he has ceased to see. But it is quite impossible that the same individual intuition should still exist when this person no longer exists. And even in thought one can keep no firm hold on this intuition without including therein the thought of the person involved. And since, now, this holds equally in relation to both parts of the intuition, it follows that, in regard to each of the two parts which I distinguish, I do not have to do with something totally other, and thus also I cannot speak of a plurality of unities here but only of the manifold character of something that is itself one. As another example one could take the case of someone who thinks a series of tones. He, too, thinks more than one successively occurring thing in one and the same moment, and he is a unity. But he is not a simple unity. He is a manifold unity, but not a plurality. We cannot call him a temporal continuum just because he thinks of something proceeding in time, for he thinks of the given succession at one and the same time. In fact we cannot count him among the continua, i.e. among the continuously many, at all, since as in the case of he who sees something spatially continuous a decomposition into totally different units and thus a conception as a true plurality is impossible.

This distinction between the continuously many and the

continuously manifold was borne insufficiently in mind by Aristotle when he inferred a spatial extension of the sensing subject from the spatial extension of the objects of sense. The consideration of what is given when someone who momentarily presents to himself a temporal continuum could have kept him from this false conclusion. It is not necessary that every part of the intuited continuum should be intuited by some corresponding part of the intuiting continuum. Indeed it can be shown that such is excluded by the unity of consciousness. If he who intuits a continuum should himself be a body, then, just as in a red surface the redness is repeated in every part and point, so the whole unified consciousness must apply to each and every part and point of the intuiting body. Aristotle however did not merely err in that he took insufficient account of the peculiarity of what is continuously manifold as opposed to what is continuously many. He also failed properly to conceive the continua themselves, in that he did not conceive them as unities which can just as well be described at the same time as pluralities. He believed, rather, that no unity can ever be a plurality, but rather that a unity is potentially a plurality and a plurality is potentially a unity.[29]

The distinction between continuously many and continuously manifold does not properly obtain in relation to what is spatially continuous as it does in relation to what is temporally continuous. The spatially continuous is always continuously many. But still, one should not of course leave out of account the peculiarity of what is temporally continuous: that it exists only in one of its boundaries. One might suppose that this would in fact exclude entirely the existence of plurality, yet this is not the case in virtue of the two-sidedness of the temporal plerosis. Moreover there comes into consideration in regard to the temporally continuous not only that which exists but also that which is to be accepted or acknowledged as factual in other temporal modes.

24. f) We shall mention only very briefly one further basis of classification which is significant in the realm of both the continuously many and the continuously manifold. This is the classification according to the genus of that which is continuous. Thus the continuous changes of time-differences and of local differences are of different genera, while the secondary local continuum, whether it exists in complete or in more or less incomplete, and in uniform or in non-uniform teleosis, is to be counted as of one genus with the primary local continuum. Further, those local continua (continua of places) which are mere

boundaries belong in a single genus with those which are not. Colour-continua, continua of temperature and so on belong to different genera, no matter whether they are given now in full, now in more or less incomplete teleosis as secondary continua, or (what stands perhaps in opposition to the now usually accepted opinion) exclusively in full teleosis. These are always secondary continua. But they distinguish themselves by the fact that they are now temporally, now spatially continuous, resting in the latter case on time as primary continuum. The perseverance of every substance and of every property of different genera, too, yields a continuum of a different genus. So also our mind in its perseverance appears to be a secondary continuum of a spatial genus. And of course that temporal continuum in which, though transcendent to us, we may grasp the life of God, is a continuum *sui generis* and one that is primary.

25. This is the place where we should speak of what one is commonly in the habit of calling the homogeneity of a continuum, whereby one is concerned in particular with the question of whether space is a homogeneous continuum. If, now, homogeneity is to be only a matter of what we just referred to as the genus of a continuum, then it would be absurd to speak of a continuum or of something continuously manifold that was not homogeneous. However, one is thinking here of various other regularities which, as we have seen already, do not apply to everything that is continuous. A line which changes its direction, and especially if it does so in irregular fashion, would therefore not be a homogeneous continuum, and nor, either, would a movement in space which varied irregularly in its direction or velocity. We have to do here essentially with peculiarities of teleosis. In looking back on what has been said earlier we shall easily see that the question of homogeneity, when raised in relation to times as such or to geometrical bodies, must be answered with a decisive yes. The primary character of these continua provides us with complete certainty in this regard.

26. g) Another easily understood classification is that in relation to magnitude. We said already that, in contrast to what is commonly believed, there obtains a relation of magnitude between every continuum and every other. This is not to say, however, that we are able in every case to work out what this relation is. The assertion loses its air of paradox when one has once convinced oneself that just as one can speak of concrete and abstract numbers so also one can speak of concrete and abstract continuous

magnitudes. If one says of a square area of side *ab* that its size is $(ab)^2$, then one may not — as did Helmholtz in an unguarded moment — conceive matters in such a way that the concrete magnitude *ab* would be multiplied by the concrete magnitude *ab*. Concrete magnitudes cannot be multiplied; rather, as in the case of '3 times 3 apples', one multiplies the concrete number by that abstract number which is equal to it. Thus here, too, one multiplies the concrete continuous magnitude by the abstract magnitude that is equal to it. I have shown already that one can in consequence say not merely that there exist lines of equal length, but also that certain times must be exactly equal in length to certain lines. It is clear that one can therefore say also that the magnitude of a certain time must be exactly equal to the square root of the magnitude of a certain area, and so on. It is a consequence of this observation that there could be no two continua, however different they might be in genus, which would not stand to each other in some relation of magnitude. Indeed there is a certain sense in which even the continuously many stands in a relation of magnitude to the continuously manifold. We speak of seven men and of seven properties of a man and can in a certain sense affirm that we are here considering men and properties of equal number. And yet the men are a plurality, the properties merely manifold. We could express ourselves quite similarly in relation to continuous magnitudes.

I have also drawn attention already to the fact that the magnitude of secondary continua always corresponds exactly to that of the relevant primary continuum. The secondary local continuum of a movement is to be assessed as to its magnitude not according to the distance of its starting point from its finishing point in space, but according to the length of time it has taken. And so movement itself is to be set as completely equal to the length of a period of rest which existed during the same time.

If instead of magnitudes of continua one is to speak of magnitudes of distances, then this is not to apply the same concept of magnitude; yet the homonymy is not accidental. It is rather a homonymy by way of relation, as obtains between health in the proper sense and the healthfulness that one ascribes to a particular region. It is not the distance of one point from another that has a length and allows the distinguishing of parts, but rather the straight line that is to be conceived as lying between them. The distance of the starting point from the end point of a rectilinear motion corresponds in magnitude to the spatial stretch described, but it corresponds to the magnitude of the relevant temporal stretch only when one takes account of the teleosis in which the different local determinations apply successively to the moving body. Should this teleosis be the same from beginning to end but twice as large in one of two motions which pass along the same spatial stretch, then for the same distance of the end points the length of this motion would be twice as large.

This modification of the concept of magnitude when one speaks of distances does not prevent us from speaking, in just this modified sense, of distances which are doubled, trebled, etc. And thus a distance can be multiplied by an abstract continuous magnitude. The distance of the direction of a given circle from that of the tangent at the point of contact is, when multiplied by the magnitude of one fourth part of the circumference, equal to the distance in direction of one right angle. Even the distance in quality between two colours, for example pure red and pure blue, must stand in a quite determinate relation to every other distance, for example to that of two tones, or even of two directions of lines or of two spatial or temporal points, a relation which we however — for very understandable reasons — are not able to determine. But still, one thing is clear: if the transition from pure red to pure blue can be executed continuously in time in every conceivable degree of teleosis, then among them must be included that degree of teleosis wahich belongs intrinsically to the primary continuum of time. And it is also clear that if a violet of just this degree of teleosis should reach red from blue in regular progression in a certain time, then the qualitative distance from blue to red would have to correspond exactly in its size to the temporal distance involved, that is to the distance between the starting point and the end point of the time taken.

Just as continuous magnitudes can stand to each other in manifold relations of size, so also they can stand in relations of shape and position. It seems however to be unnecessary to go into these in depth.

27. We prefer to turn our eyes at the close of this investigation to the question of the empirical or a priori character of geometry. It is obvious that, like all our presentations (even those of number), our geometrical presentations, too, have an empirical origin. And if it is our knowledge that there exists a space of three dimensions that is at issue then it is indubitably empirical, and is identical — for all that has been said against this idea — with the knowledge that bodies exist. The question as to the a priori character of geometrical knowledge

can therefore amount only to the question whether, granted the relevant presentations and the knowledge of the existence of bodies, the axioms follow from the analysis of these presentations and thus are all cases of the law of non-contradiction. In the light of what has been set out above, this question must be answered with a decisive yes. For as certain as it is that a body is not the mere boundary of a presently existing four or more dimensional continuum, so it is certain that it must be what has been called a plane body. And as certain as it is that it is a primary continuum, it is certain also that no irregularities of teleosis can exist within it. It is, therefore, what one calls homogeneous. More specifically in relation to the axiom that between two points there is only one straight line, it cannot be doubted that, if the line is thought of as in full plerosis, then the evidence for the axiom is seen to lie in the very presentations, and there is just as little room for doubt in relation to the much disputed 11th axiom of Euclid. For it is certain that the sum of the angles whose vertices meet in a point of a plane is equal to four right angles, and in the same way it is certain also that the sum of the three external angles of a triangle is this too. And since these form, with the three angles of the triangle, three pairs of adjacent angles, and thus have six as their sum, so it is certain that the sum of the angles of the triangle must be equal to six right angles minus four, that is to two right angles, which establishes also the truth of the 11th Euclidean axiom. It is beyond doubt that geometry can take account also of hypothetically assumed topoids of more than three dimensions, and that it can deal with non-plane three-dimensional boundaries which such topoids can be conceived to possess. But far from its being the case that the 11th axiom will be upset thereby, it will much rather be called in aid in the calculations which then take place.

Addendum to the treatise on what is continuous

Dictated 16 December 1915

1. In relation to the question as to the origin of the concept of continuity, we fought those mathematicians who want to arrive at this concept not through abstraction from intuitions but through induction. We paid particular attention to Poincaré, who follows extreme empiricists in the area of sensory psychology and therefore does not believe that there is granted to us an intuition of a continuous space. Poincaré's entire mode of procedure reveals that he also denies that we are in possession of an intuition of a continuous time. We saw how he first of all inserted between 0 and 1 fractions having a whole number as numerator and a whole power of 2 as denominator. In similar fashion, he then inserted all proper fractions whose denominator is a whole power of 3, and then also those whose denominators are whole powers of every other whole number. He obtained thereby a series consisting of all rational fractions which, as he said, already has a certain continuity about it. He then inserted all proper fractions whose denominator is the root of such numbers as are not merely the results of the relevant multiplications of the numbers by themselves, that is a series of irrational fractions. To these were then added also a second class of irrational fractions which present themselves as having more complex demominators, as for example $1/(\sqrt{2} + \sqrt[3]{2})$, where, however, the denominator is still always to be expressed by means of numbers and function-signs. He held that through intercalations of this sort one arrives at a series of still greater continuity. To these one now adds the series of fractions involving transcendental ratios (as, say, between the radius and circumference of a circle). Poincaré was prepared to admit that this process will never actually come to an end, since after one has taken account of one or more transcendental ratios there will always remain others not considered. But he believed that he could be satisfied with the insertions already made. And nothing is more self-evident than that we have here a confession that the attempt to obtain a true continuum in this way has broken down.

2. It is not merely the demonstration of gaps still remaining which makes this clear, but also the fact that the concept of a

transcendental ratio cannot itself be thought without appeal to the concept of the continuum. It is after all a relation of such a kind that it cannot be expressed through numbers and functions thereof. Thus π, which represents the ratio of circumference to diameter, allows of no corresponding numerical expression. The infinite decimal expansion 3.14159... is merely an arbitrarily close approximation. Hence the concept of the continuum must here be brought in aid — but this concept is supposed to be not yet in our possession.

3. Already in relation to the algebraic irrational fractions the defective character of the whole enterprise is clear. For after all every proper fraction is, like every negative number, an absurd fiction which can be admitted without demur only because to the fraction one can add a designation possessing a divisibility which the numerator of the fraction lacks, as for example in the case: half a dozen. In the case of rational fractions this designation can be a number; in the case of irrational fractions however it can only be a continuum. But since I am attempting to construct the continuum, I still lack anything to specify the fraction, and this is the only thing that would allow the irrational fraction to be admitted in spite of its absurdity. Thus here already there becomes manifested that viciously circular procedure which we have encountered still more palpably in our consideration of transcendental irrational fractions above.

4. Poincaré, in his attempt at construction, has departed at certain points from Dedekind who counts as the first to have succeeded in such an attempt. Dedekind differs from Poincaré already in the fact that he does not wish to deny that we have an intuition of a continuum — he simply does not want to make any use thereof. Moreover he concerns himself not at all with the transcendental irrational ratios. (This is certainly not an advantage, if it serves to prevent the breakdown of the attempt from manifesting itself so clearly.)

5. Dedekind's and Poincaré's constructions share in common that they fail to recognise the essential character of the continuum, namely that it allows the distinguishing of boundaries, which are nothing in themselves, but yet in conjunction make a contribution to the continuum. Dedekind believes that either the number 1/2 forms the beginning of the series 1/2 to 1, so that the series 0 to 1/2 would thereby be spared a final member, i.e. an end point which would belong to it; or conversely. But this is not how things are in the case of a true continuum. Much rather is it the case that, when

one divides a line, every part has a starting point, but in half plerosis. Euclid's supposition that a point is that which has no parts was seen already by Galileo to be in error when he drew attention to the fact that the mid-point of a circle allows the distinction of just as many parts as there are points on the circumference, since it differs in a certain sense as starting point of the different individual radii. If a red and a blue surface are in contact with each other then a red and a blue line coincide, each with different plerosis. And if a circular area is made up of three sectors, a red, a blue and a yellow, then the mid-point is a whole which consists to an equal extent of a red, a blue and a yellow part. According to Dedekind this point would belong to just one of the three colour-segments, and we should even have to say that it could be separated from this while the segment in question remained otherwise unchanged. Indeed the whole circular surface would then be conceivable as having been deprived only of its mid-point, like Dedekind's number-series from which only the number 1/2 has completely fallen away. One sees immediately that this is absurd if one keeps in mind that the true concept of the continuum is obtained through abstraction from an intuition, and thus also that the entire construction has missed its target.

6. To make this clear by means of yet another consideration we shall for a moment make the (in itself actually absurd) assumption that infinite space would contain at one and the same time a collection of spheres, each moving with a different velocity. For one sphere this would be 0, for another 1 mile per hour, for a third 1/2 a mile per hour, and so on, so that there would be represented by some sphere every intermediate velocity between 0 and 1 mile per hour that is conceivable at all, whether it manifests a rational or an irrational ratio. If one asks whether one would then have to do with a continuum of velocities, then this question would, according to Dedekind, have to be answered in the affirmative. In truth however it would have to be denied. Where an actual continuum of velocities would be present is in the case of a disc rotating in such a way that the velocity at the circumference is 1 mile per hour while the centre did not change its place. The difference between the two cases is this: in the latter, each of the velocities appears as a boundary which taken in itself is nothing, but when unified with the continuum of velocities is such as to make a contribution thereto; in the case of the collection of spheres, in contrast, the velocity of each sphere is something for itself; it is just this which stands in

contradiction with its forming a true continuum with the remaining velocities.

7. Even though the existence of an actual continuum (for example of a spatially extended body) is not immediately evident, the proof that we possess intuitions of continua is of great significance in settling the question whether the concept of the continuum is subject to contradictions. It is well-known that there is much that can lead one to the temptation to declare it as contradictory. The *solvitur ambulando* is not here applicable, since it is the actual *ambulari* that is called into question. However, the intuition of a continuum must at least then be admitted if the concept is gained by abstraction therefrom, since that which is contradictory cannot merely not *be*: it cannot be intuitively presented, either. Thus for example, one cannot present a red white horse, yet we can perfectly well acquire the concept of a red white horse and other absurd concepts through composition and attributive connection. Thus we have a very simple and compelling refutation of the asserted impossibility of a continuum which would not be available to us in the case that this concept were the result of a construction.

Concluding remarks

8. To be clear about the total failure to reach their target of Dedekind's attempted construction and of those of all other mathematicians who want to obtain a continuum through intercalation of fractions, it seems useful to consider the following. Just as it belongs to the nature of a continuum that its parts are in contact with each other as boundaries, so it also belongs to its nature that it has a degree of variation. In the case of time, and of space as this exists simultaneously in all its parts, this degree is constant, and that is why the distance between two points in time corresponds exactly to the length of the time proceeding between them and why the distance of two spatial points corresponds exactly to the length of the straight line that lies between them. In the case of other continua in contrast, for example movement, it is variable. Different movements have very different velocities and even in the case of an uninterrupted movement the parts can still vary in multifariously different degrees. Thus it can be that now a

42

very long, now a very short movement along a straight line may lead from one identical point to another, and the size of the distance between the two points can stand in arbitrary relation to the size of the movement.

If, now, one could really construct a continuum between 0 and 1 by an intercalation of fractions in the way the mathematicians want, then clearly this continuum too, would have to possess a certain degree of variation, and this would have to be conceived either as constant and absolutely invariable, as in the case of space and time, or as arbitrarily variable, as in the case of movement. In the former case the length of the continuum would correspond exactly to the size of the distance between 0 and 1. In the latter, in contrast, one could not speak of any fixed relation at all. Strangely enough, the mathematicians have neglected entirely to address this problem. If however the question were put to them, what would they answer? I believe that there can be no doubt about this. They will have to answer that no fixed degree of variation is called for, but that every arbitrary degree is equally conceivable. One could indeed, if a complete enumeration of all the fractions between 0 and 1 were only possible, go from 0 to 1 via an enumeration of arbitrary degrees of velocity, just as a movement can lead with arbitrary velocity from one place to another. If we now however examine the general conditions for such a continuum of arbitrarily changing degrees of variation, then it becomes manifest that it must be a secondary continuum, i.e. that another continuum must underlie it which has a necessarily constant degree of variation and is therefore such that, if it does not vary in its direction, its length corresponds exactly to the magnitude of the distance between its end points.

From this it follows that the number-continuum which is supposed to have been produced would be in every case a secondary continuum which therefore included the idea of a primary continuum as previously given. And thus for this reason, too, it is apparent that what one supposed oneself able to gain by construction has already — without this being noticed — been presupposed as given. The contradictoriness of the whole enterprise thus comes to light once more in the most flagrant manner.

Someone might want to assert that in his assembling together of all the numbers between 0 and 1 he does not think of an enumeration but of a simple co-existence of all the given fractions. But then one should have to ask him whether the fractions thereby co-existing would not have to manifest a certain density, and whether this would have a constant degree which would be from the very start the only degree conceivable. And if so which? In the case of constantly accelerating motion, or of a grey surface which passes in constantly changing degree of variation from a purely black to a purely white line, all degrees of variation are able somewhere to be represented. Hence, if it were evident that in the case of a continuous number-transition only *one* degree of density were conceivable, one would have to be able to say which degree of variation appropriate to some temporal point of the motion or to some spatial line of the grey surface would be that of the continuum constructed out of fractions. It is clear that no decision in regard to this question would allow even the appearance of rational justification.

Thus the error of a *petitio principii* has in various ways been demonstrated for the whole attempt. The continuum that is supposed to have been gained is devoid of boundaries and of every degree of variation.

II. On the measure of what is continuous

1. Every continuum is infinitely differentiated, and therefore homogeneous.

2. The same thing can fall under several genera, and therefore two things differentiated into species in merely *one* genus can be the same in all others. And the parts of a continuum, while being continuously differentiated in one genus alone, can be the same in others, or manifest a finite number of differences. Examples: A lasts for a certain time. A and B, succeeding each other, last together for one hour.

3. It can also happen however that the parts of a continuum are continuously differentiated in *several* genera. I mean here not the sort of case which presents itself as a continuum of several genera in such a way that in the one there is a differentiation exclusively in this genus, in the other exclusively in that, as for example when a sphere is at rest for one hour. In this case of a sphere at rest we have something spatio-temporal in which the upper hemisphere is distinguished spatially from the lower, while that which exists in the first half hour is distinguished temporally from that which exists in the second, and so on *in infinitum*. I mean, rather, that which presents itself as a continuum of several genera in such a way that the same parts are differentiated in two or more genera — cases such as that of the movement of a point conceived as indefinitely small proceeding from *a* at time α to *b* at time β.

4. Only this latter case manifests an analogy with the case of successive existence of A and B where what exists is differentiated both temporally and qualitatively, though in the one respect continuously, in the other discretely. The former case, in contrast, would have to be compared with the case where A and B exist together for a certain period of time.

5. It is for this reason that we would have in the former case an increase of the continuum, as in the case of unified existence of A and B. Where, in the case of unified existence, we should have to execute an addition to determine the total magnitude of the spatio-temporal continuum involved, in the case of that which is doubly continuous a multiplication would be necessary. In the case which interests us, however, such a multiplication would be simply

out of place. And quite generally one could not speak of an increase in the continuous magnitude through the adjoining of one continuum onto another. One need only compare the case of successive existence of A and B. The continuum here does not undergo a doubling of size, as in the case of the simultaneous existence of the two.

6. But what will this magnitude be, that of the space through which the motion proceeds, or that of the time the motion takes? Already the analogy with the case of the discrete differentiation of A from B speaks for the idea that it will be that of the time. This is confirmed by the fact that the time appears to be filled out with equal completeness in the case of rest as of motion, and in the case of quicker as of slower motion. Every spatial determination, in contrast, applies the more incompletely to that which it localises the more rapid is the latter's motion. Imagine two concentric circles having radii r and $R = 2r$, which are such that they make arcs ab and ab of equal length on their respective circumferences. If the plane is now turned about the centre, so that one complete rotation is effected in an hour, then in this hour $a b$ has traversed the entire smaller circumference while its equal, $a'b'$, has traversed the entire larger circumference. And this has taken place in such a way that when ab occupies the nth part of the smaller surface, $a'b'$ occupies the $2n$th part of the larger surface.

With Aristotle we want to call these differences in completeness differences in the *teleosis* of spatial determination (localisation). Since in a case of increasing velocity the spaces described increase to just the extent in which the completeness of the participation in the spatial determination is decreased, it follows that the one is compensated by the other, and we end up always with the same measure of differences both in relation to space (to motion) as also in relation to time; we end up, that is to say, with the doubly, spatially as well as temporally, differentiated collection.

7. In the double continuum which we have before us in the case of a movement, we call the continuity of temporal differences the primary continuum, that of the successive differences of place its secondary continuity. This can itself be manifold, so that the double continuum comes to be a triple continuum. This can occur in so far as the velocity, instead of being uniform, is differentiated and perhaps continuously differentiated — which is itself possible either in a uniform or in a differentiated fashion. Or it can occur in so far as the direction of motion, instead of being the same, is

differentiated and perhaps continuously differentiated, as in the case of motion along a curve, etc.

8. Even a line given as a simultaneous juxtaposition (a spatial length of indefinitely small width and depth) is a double continuum. Here the primary continuum is the length in space, the secondary is the curvature. Just as the measure of the velocity of a motion can vary, so also can the measure of the curvature of a line — in the case of a circle, halving the radius leads to a measure of curvature that is twice as large. The curved line manifests continuously differentiated directions in incomplete teleosis, while the tangents manifest these directions in complete teleosis. The spatial length is the magnitude of the continuum and this is the same whether we are dealing with strongly or weakly curved or with straight lines, though in the one case there is only one direction, in the other cases many directions and a greater or lesser collection thereof, but the greater this is, the less is the average teleosis.

9. One sees easily that in order that there be a double continuum, continous differences must be possible which belong to two different primary continua or to a primary continuum of more than one dimension: in the case of motion, space and time; in the case of the curved line, space of more than one dimension.

10. While the teleosis of the secondary continuum can be manifold, that of the primary continuum is always and everywhere the same. Where the velocity of differentiation in the former case can be increased to infinity, and, when increased to infinity, can arbitrarily approximate full teleosis, the primary continuum always retains one and the same measure of succession of differentiation no matter how far the continuum extends. It is clear that an undifferentiated continuum can nowhere take the place of a differentiated one. Everything is in a process of transition to something differing in species. But if there is thus no realisation of any single species in full teleosis, so there is also no possibility of falling away from that teleosis which is common to and necessarily characteristic of all primary continua. We call this the *primary teleosis*. It is necessarily equal to some one of the teleoses to be found among secondary continua.

11. Space and time, both primary continua, therefore have the same teleosis, though not homogeneously. A still greater uniformity is manifested, in virtue of their homogeneity, by times compared with times, spaces compared with spaces.

That which is temporal and that which is spatial always have dimensions and each has always the same number of dimensions. There can be no temporal point in the sense of an extensionless moment, and no spatial point, line or surface, except in the sense of dimensions that have become indefinite and so to speak shrunken in a certain direction.

Not merely differences in just as many directions are possible for any point, but also in the very same direction. However small a continuum may be, the distances and connections in all directions possible within the continuum are always actually given.

Part Two

TIME AND TIME-CONSCIOUSNESS

I. What the philosophers have taught about time

Before 1902[30] *[T 4]*

1. *Introduction.* What is time? There is no other name that is more familiar to us, and none that is at the same time so obscure. Wherever we employ it in speech no difficulty accrues to our understanding and we are also able to tell easily and surely whether a particular determination is temporal or not. And yet many will hesitate in giving an answer to our question and perhaps in the end admit, like Augustine in a famous passage of his *Confessions*, that they do not know. The one is entirely compatible with the other, just as schoolchildren when they learn their ABC can understand what is being spoken of even though they could gain insight into the nature of each letter only on being introduced to the Helmholtzian analysis of sound.

The practitioners of mechanics, for whom the concept of time belongs to the first and most important concepts of their science, have allowed themselves to be satisfied with this general comprehensibility of the expression. Neither Archimedes nor Galileo felt the need to embark on investigations into the nature of time, though psychologists and metaphysicians have repeatedly concerned themselves with the clarification of this riddlesome concept in extensive discussions.

The metaphysicians! I need only utter the word in order to awaken the most unfavourable of prejudices, given the contempt with which this once so celebrated discipline is regarded today. If an investigation is metaphysical, one will say, then it is already marked out as pointless. And if we look at history, then the inadequacy of the definitions, which makes almost every greater thinker contradict his predecessors and even nowadays does not allow agreement, seems little suited to dampen this mistrust.

2. *Aristotle.* Aristotle gives the following explanation of time: it is the number of the movement of the uppermost celestial sphere in

so far as this supplies the measure of the earlier and later for all other change and perseverance. If one sets aside the errors attached to this definition as a result of its connection to the pre-Copernican astronomy, then it asserts nothing other than that the question 'When?' is customarily answered by determinations which relate to the changing state of the heavenly bodies. Aristotle proceeds here in a manner similar to the way he proceeds when he designates space (υοποτ) as the proximate stationary boundary of the *surroundings*, because we customarily respond to the question 'Where?' with relative determinations starting from a stationary boundary (at least in relation to that part of the corporeal world which concerns us most). Where is he? In the room. Where is the room? In the house. Where is the house? In the city of Florence. And this is in the province of Tuscany and this in the Kingdom of Italy.

To believe that the Aristotelian definition gives us insight into the nature of time would be just as ridiculous as if someone were to say that the essence of warmth consists in the thermometer in so far as this provides the measure for the more and less of temperature. Aristotle has not so much defined the essence of space and time as named that to which, because we presuppose it as sufficiently familiar in its spatial and temporal determinations, further relative spatial and temporal determinations are referred.

At another point he says of time that if there were no soul then there would also be no time, and many were not ill-pleased to call him in aid in this connection as a predecessor of Kant with his doctrine that time is a subjective a priori form. But clearly the latter proposition is nothing other than a consequence of the former. If there were no thinking thing, then certainly, according to Aristotle, the motion of the uppermost celestial sphere would continue. But it would then no longer serve as measure of remaining or changing in regard to its earlier and later. For the question what time is, there is in practice only *one* of his utterances that is of a certain significance, though time is not in fact mentioned therein. I mean the passage in the books *On the Soul* where, in treating of the objects of sense, he distinguishes between objects of a sense that are proper to it (the sensory qualities) and objects of sense that are had in common, under which he mentions, among others, motion and rest. Since these latter phenomena are not to be thought without a before and after, it follows that Aristotle apparently believes that there is something given in our original intuition that could serve as starting point for a clarification of the concept of time. He did not however provide the analysis that would yield such clarification.

3. *Augustine.* When we turn from the greatest thinker of pagan antiquity to the greatest thinker of the patristic era, then we find that Augustine approaches the question of the essence of time with genuinely philosophical wonderment. He relates that a learned man had explained to him that time was the motion of the stars. It is not difficult for him to show why this definition, which is visibly related to the Aristotelian, is unacceptable. But he is less successful when he attempts a solution to the problem of his own.

It appears to him as indubitable that time is an extension and that it is measurable. On inquiring further, however, what it would be the extension of, he comes to a result which can be summarised briefly as follows: it is an extension which applies to the created things in so far as they appear before our created mind as something which it looks back to as past, grasps as present, or looks forward to as future. In and for themselves, he holds, and thus also in the knowledge of God, perishable things do not have an extension of this sort. Otherwise, he asserts, God would not be free of change, which appears to him as quite simply unacceptable. It follows from this that we must in fact characterise Augustine as a representative of subjectivism and of the purely phenomenal truth of time, though then there clearly stands in sharp contradiction to this that he also speaks of things in themselves as subject to change and as perishable.

4. *Aquinas.* The lively interest which the great Father of the Church showed for our question and the dissenting criticism which he directed in this regard towards Aristotle, the supreme philosophical authority of the middle ages, could not but have a stimulating effect upon the scholastics. According to Aquinas one should hold on to the idea that time is the number of the measure of motion in regard to earlier and later. But it becomes clear that he understands by 'motion' every kind of succession, for which not one but several measures are allowed to hold. He distinguishes in particular between inner and outer time. An inner time is every succession in so far as an ordering of earlier and later is to be found therein — of which there are therefore many. As external measure in contrast he adopts, at least for all corporeal motions, a unified standard, and indeed this is to be, as in Aristotle's case, the motion of the uppermost celestial sphere. Even though here, as elsewhere, his dependence upon the great Greek is clear, still certain innovations by which he went beyond Aristotle are likely to have

been stimulated by Augustine's criticism. These are, it must be admitted, marked by very dubious consequences, some of which were already expressly drawn by Thomas himself. Thus since changes are so various, times too are supposed to be very different in their nature. This leads him so far as to assert of the angels that their time would be not a continuum but rather, because their thought changes in a discrete series of jerks, a discretum of infinitely many unextended phases.

Yet since even the movements of bodies are such as to be in part coming to a complete standstill, in part irregularly slowing down or accelerating, he ought to have adopted also here no less great and peculiar variants of time; he did not however see this clearly. For God, however, in whose activity, like the Church Father before him, he denies all succession, he does not want to accept anything analogous to our time at all.

Such a doctrine palpably fails to do justice to the nature of time. However difficult it may be to give a clear account of the concept of time, still there are certain determinations which it undeniably has according to the judgment of every healthy intellect. Thus it is not merely that there must be an earlier and a later wherever there is a time — this is acknowledged also by Thomas — but also that no differences in velocity can occur in the passage of time. Every physicist makes use of this insight when he expresses velocity by means of the formula $v = s/t$. If time itself could pass now more quickly, now more slowly, then this formula would lose all determinateness.

5. *Suarez*. Francisco Suarez, a thinker held in high esteem by Grotius and Leibniz, who shares in essentials the views of Thomas, is not afraid to draw also the following absurd consequences. If two angels are created simultaneously and the one is annihilated after one year, the other after one hundred years (or not at all), then the one would exist, as far as concerns its internal duration, for just as long as the other. It is only the external measure of their durations which would be different, in so far as the one would co-exist with a longer, the other with a shorter motion (*Disputationes metaphysicae*, 50, sect.5). Similarly he asserts without reservation that if the one had been annihilated and then recreated anew, then the duration which it would acquire after the second creation would not extend its total duration but would rather be individually the same as its original duration. Finally he frankly declares that every single time could return arbitrarily often as individually the same, since one and the same motion can be repeated arbitrarily many

times. Thus individually the same day as is now coming to an end can begin again anew. Indeed it seems to me that Suarez would consistently have to say that this in fact happens as often as day follows day, and that he would have to say of every single hour in the day that it was individually the same, before and after every 'other'.

6. *Bonaventura.* Practically all scholastics are in virtual agreement with the strange conception of time, which underlies all of this.[31] Only Bonaventura and a small number of thinkers influenced by him constitute an exception, and reveal an essentially different conception. According to Bonaventura, everything other than God (who is not temporal but eternal) is subject to continual change. Even should something created continue to exist unchanged, it would still suffer a continual alteration in its being, since it is incessantly sustained, i.e. produced anew, by God. According to whether these uninterruptedly continuing sustainings form a longer or a shorter chain and end earlier or later the duration in time and temporal location of the thing are different.

Many of the absurdities of the more usual medieval conception of time and duration are without question avoided by Bonaventura. But *one* difficulty remains, as a result of which this attempt, too, must fail. God is supposed to be unchangeable. But now even if something which proceeds from him is unchanged in respect of what and how it is, how is it possible that the emission of this unchanged something from an unchanged God should be continuously something new? In all reason one could answer only this: that even if they are in all other respects the same, still the one emission is different from every other in regard to its time and it is for this reason that the one is earlier and the other later and that there occurs a shorter or longer succession of such an earlier and later, i.e. precisely a difference of duration. But now if one would adopt this, then the whole explanation would clearly have come to be an *idem per idem*[32] and thus it would be clear that the concept of time has been clarified not in the slightest. And it would appear inexplicable, too, how, if time were a succession of processes[33] lying entirely beyond our intuition, we could have quite adequate knowledge thereof without the need for much reflection, so that we can all quite easily grasp enough concerning the most varied determinations of time as to be able to employ them in everyday speech.

Thus the endeavours of the scholastics reveal themselves as completely inadequate.

But one thing must be noted before we take our leave of the middle ages. The conception by which Augustine felt himself tempted, without however carrying it out consistently, namely the conception of time as something purely subjective, was not revived by any of the scholastics. Quite the contrary: we see them concerned to demonstrate, by exploiting the Aristotelian doctrine of motion as *actus imperfectus*, the logical possibility of those paradoxes which had pointed Aristotle in the direction of such a conception.

7. *Descartes.* If we turn to the modern philosophers, then according to Descartes time is an attribute of substance. It is the same as what is also referred to as duration. Duration is remaining in being. It can be exactly the same in case of both rest and motion, otherwise it would have to be impossible that one body is at rest and another is moving in the same time. Yet we measure the duration of something by taking as standard the duration of a motion that is proceeding with maximum uniformity. And the determination of time is that determination of relative magnitude which applies to a duration when we measure it against a common standard of this sort. Thus time and duration are in themselves identical and differ only in our conception.

There is much here to remind us of the definition of the scholastics. Much that is erroneous has been cast aside. But one thing is still not clearly expressed, and that is whether Descartes conceives the perseverance of a thing at rest as being without any inner change and without any succession of different parts. Suppose such change or succession were missing. How, then, could a continuous extension be arrived at? Suppose such change occurred. What, then, is it, which follows what? Being? Does one part of being follow another part? If the parts are to constitute a plurality, they would have to be different, but in what do they differ?[34] Descartes counts duration as belonging to that in the things which manifests itself to us. But he makes insufficiently clear wherein the inner differences lie that we are supposed to find here. Are they *being past*, *being present*, and *being future*? And in regard to being past, are there degrees of being past, etc.? And is being past really in the enduring thing as is being present? Augustine, at least, would have grave misgivings in accepting this. But we cannot even say with certainty whether Descartes wanted to assert this or something else. Even the manner in which the concepts of duration and time are set in relation to one another seems dubious. Descartes treats them as if they would relate to each other in a way similar to magnitude and a measure of magnitude, though in fact they seem much rather to be related as volume to place.[35]

8. *Locke.* If Descartes' explanations still recall something of the middle ages, then this can no longer be said of those of Locke. The whole tenor of his famous *Essay concerning Human Understanding* implies that his determination of the concept of time is related most closely to the question as to the origin of the relevant concepts.

We properly receive the idea of succession from the senses. Yet it is presented to us more constantly by inner events (Bk.II, Ch.7, §9). Everyone grasps with evidence, as soon as he reflects upon himself, a series of ideas which follow one another constantly for as long as he is awake. This gives us the idea of succession (Ch.14, §3). We also notice distances between the parts of this series, and this leads to the concept of duration. For while we are receiving a number of ideas one after another, we recognise that we continue to exist, and our continued existence is commensurable with this succession. The succession of ideas can manifest only slight differences. (Given greater speed the succeeding impressions would become indistinguishable; given greater slowness they would no longer be together in consciousness.) That part of duration which is just equal to one of the ideas following another within us is called an instant. A constant and regular succession of ideas in the waking man becomes thereby the measure of all other successions. By means of the senses we observe certain phenomena which appear as set equally apart in their succession. Thus we distinguish regular periods of determinate lengths of duration: minutes, hours, days, years and so on. Our imagination thereby exceeds what we perceive, by duplicating it arbitrarily. Thus we come to the idea of a tomorrow, of a next year, of the next seven years which are to follow the present moment (Ch.14, §31), and further still to the idea of a duration without beginning and without end, i.e. of an eternity. Time is however nothing other than a part of this infinite duration when the latter is conceived as divided into measured periods (loc.cit.).

Elsewhere he refers to time as the whole of duration as set out by certain epochs of equal measure (Ch.14, §17). And in yet another place he says that time is to duration what place is to spatial extension. They are both, he says, so much of those boundless oceans of eternity and immensity (Ch.15, §5).

In appealing to certain intuitions in his explanation of the concept of time, intuitions in which this concept concretely appears, Locke testifies to a very correct methodological conviction. But one cannot praise him for having shown skill in the exploitation of these intuitions.[36] That time is a matter of succession had been said already by Descartes and by many before him. He had left us in the dark only about what it is that succeeds what in time,[37] and about what the specific differences are which the distances of different magnitudes entail. But he thereby left us in the dark also about the whole concept of time, just as we would remain in the dark about the concept of colour if we were left unclear about the genus of those specific differences which make up the distance of red from blue or of black from white. One is sometimes almost tempted to believe that every sort of difference could constitute these distances, and indeed that they would be constituted in particular by the differences of the succeeding ideas.[38]

But how could that be, given that the same two ideas can follow each other now immediately, now mediately, now quickly, now more slowly? And with so little clarity about the idea of time it appears also quite unjustified for Locke to hold time to be continuous, especially when he admits that he is, in regard to the existence of certain ideas, able to notice no succession at all. When he refers to the time during which there would exist the individual idea in its regular flow as an 'instant', then he almost suggests that he holds time to be a discrete series of moments. In any case, this confirms still further that, as I just said, he has no right to declare time to be continuous. And he seems just as little to be justified, either, in asserting that time could be extended to form an eternity. He himself teaches elsewhere that we have no positive presentation of such an infinity. Thus here, too, he appeals only to the possibility of conceiving a given temporal duration as repeating itself in imagination. But someone could just as well hold that imagination could continue the different series of intensities in both directions to infinity, that the series of nuances of grey could become forever lighter and darker, that sounds could become higher and lower in ever new scales. And he would be simply wrong were he to deny here any boundary that were incapable of being crossed.[39]

We have seen how most of the schoolmen of the middle ages ascribed to certain things a perseverance with inner succession of differentiated parts. Others however ascribed to such things perseverance without such inner succession; and they held that because of this no length should accrue to their inner durations. Such inner durations ought, according to them, to have no lengths, which did not however prevent them from asserting, with evident absurdity, that such lengths would show themselves to be equal when measured against a succession external to the things in question which would begin and end with them. Locke, too, seems to ascribe to certain things a perseverance that is unchangeable in every respect, while succession would apply to others, as for example to the substance of our soul. Has he, too, not ascribed to such things both an absence of internal length of duration and also a simultaneous external duration which has a length? Or has he after all ascribed to them an internal length of duration but without any succession of internal parts? This would be no less absurd. If then there must be present here, too, a comparable length, and a comparable plurality of comparable parts succeeding each other and together making up this length, then what would be the nature of these parts and of their differences?[40] Clearly it is time itself that he would have been aiming at with such questions, but in failing to raise them he does not notice the extent to which he has left the concept essentially unclarified.

9. *Leibniz*. In following page for page in his *Nouveaux Essais* the inquiries of Locke, Leibniz is thereby also led to elaborate on the essence of time. And this he does by criticising the definitions given by Locke. Thus we find there the continuity of time clearly expressed and strongly emphasised. The use of the expression 'instant' for the duration of existence is rejected as more popular than scientific, and an analogue of the spatial mathematical point is substituted in its place. Time itself however is for him an analogue not of three-dimensional space but rather of a straight line. It is a completely regular succession which proceeds in no part either more quickly or more slowly. The flow of our ideas is in contrast not completely regular in speed, indeed it does not even approximate to a completely regular succession even to the extent that Locke supposed. Time is not at all to be found in the realm of the actual world. Rather, our reason is led, in view of the greater or lesser approximation to full regularity, to form the idea of a completely regular succession which then serves as standard of measurement for all actual successions. And this standard is time. Time is therefore an ideal of completely regular succession manifesting itself to us in this finished form neither to the outer nor to the inner sense. It is rather a creation of our minds.

Yet still, matters stand for Leibniz much as with the idea of space. Leibniz calls this the order of all that co-exists. Sometimes however he talks as though he is thinking here of a standard or measure that would supply a determinate position for each thing that co-exists. Many things (one can with probability say all things) are in motion, though some to a greater, others to a lesser extent. Thus we come to the idea of absolute rest. And things conceived in mere thought as absolutely at rest are then to be that in relation to which every position and every change in spatial location is to be measured. Whether we can carry out this measurement exactly matters not at all, for this is, absolutely speaking, in and for itself, the natural measure, the reference point of all order.

If we examine what Leibniz criticises in Locke here in the light of our own criticisms above, then we find that Leibniz does not touch at all some very essential points, precisely because he too has not managed to free himself from the essential defects of Locke's conception of time. Even if none of the changes which take place in the actual world proceed, as a matter of fact, with total regularity, still it is not contradictory that there should be a change of which this were true. And if a completely regular succession were somewhere given, would this therefore be a time? Clearly not, since the nature of time is such that an acceleration or deceleration in its case is simply unthinkable. If now, however, in relation to the changes actually given in experience, we conceive of a change that proceeds completely regularly, then this will clearly be of the same general nature as the given empirical changes and be conceived by us only as standing out from them as an exemplary norm. This too, therefore, cannot be time. There must, then, be some moment of the phenomenal world whose nature is such that it proceeds of necessity in a uniform way; our job is to grasp this moment through abstraction.

Consider also the following. Our idea of time is supposed to be the idea of an ideally uniform succession. What belongs to such a succession? Clearly it can suffer no sort of longer or shorter interruptions, no interruptions at all. But the absence of gaps is not sufficient. Even an uninterrupted motion can still in its individual parts be of highly irregular velocity. Would it be a time in every velocity? There would then exist quicker and slower times. Would it be time in only *one* of its velocities? But then why precisely in this one? Surely not simply because it is regular.

One sees from all of this that time's being set apart from other changes must lie in some respect other than that of complete regularity. And we see once more that Leibniz has left an essential hole in his account which is the same as that which we complained about in Locke. He does not give the nature of the specific differences which determine distances in time, and thus also he leaves the whole concept of time in a state of unclarity.

58

10. *Hume*. David Hume, too, criticised Locke in a number of important points; for example he uncovered errors in his treatment of the concept of cause, errors which he then — admittedly not very successfully[41] — tried to correct. In Locke's treatment of time, however, he finds nothing to object to. He accepts that we perceive, if not a causation of events, still, certainly, a succession. That which would be worth our notice in such a perception is however discussed by him no further, and he is clearly satisfied with what Locke had had to say.

11. *Kant*. Kant, in contrast, sets himself up in confrontation with both Locke and Leibniz, essentially disapproving of both.

This philosopher, as is well known, places great difficulties in the way of understanding,[42] and even today, after 100 years, his commentators can still not agree amongst themselves. This probably rests in part on his clumsy style and on the plethora of new terminology, much of which, given greater elegance of expression, he could have done without. For the main part, however, it rests on the fact that, for all the significant power of mind revealed in the daring of his constructions and systematisa-tions, Kant possessed only a modest capacity for psychological observation. To overlook this, one has to be struck with blindness of a sort which is, however, not rare among his admirers, in consequence of which they are still less capable of observing the idiosyncrasies of their master than he was of observing the psychological facts. If now, as is natural when studying an author, one looks not merely at his works but at the same time at the objects with which they deal, and then finds nothing to which the words could be applied, then one starts to have doubts about their sense and to make attempts to change them around, this way and that, which then leads to the hundred differences of opinion and finally to general uncertainty.

What I say holds, too, of the Transcendental Aesthetic, where Kant sets out his conception of space and time. In this respect it is useful to us that we come to Kant from the consideration of Locke and Leibniz, with whom Kant occupied himself for several years and under whose influence he stood before writing his *Critique of Pure Reason*. For paying attention to the doctrines of predecessors is often no less useful to understanding than is looking at the objects themselves.

Locke, like Leibniz, had sought in time a principle of the order of successions, a unitary measure for determining the distances between what is earlier and what is later. Locke believed this to reside, in the last analysis, in the flow of ideas, which is supposed to proceed with virtual regularity. Leibniz, in denying this, substituted for it an ideal which, since there is nothing completely regular outside the mind that would correspond to it, is supposed to be the product of our conceptual thinking. Both equally conceived it as something that is extrapolated by us *in infinitum* towards both the earlier and the later. Time is thereby, according to Leibniz, continuously one-dimensional, comparable to an infinite straight line. His conception of time is however like that of Locke subject to the severe difficulty that even a completely regular and continuous succession can still be subject to the greatest of differences. Thus a completely continuous and uniformly proceeding transition from blue to red through the transition-phases of violet (blue-red) can take place now half, now twice as quickly. An ordering in relation to these phases[43] could therefore not yet yield any final determination of the relevant temporal distances, since it itself would still have to be measured in regard to its velocity. Consider a repeated transition from blue to red and back again. Each transition might take place quite regularly and yet each in a different time,[44] although the same colour-intervals would be connected together. The transitions are therefore in time and presuppose it; they are not themselves time.

What, now, does Kant do? On the one hand he holds on to the idea that time is a unitary measure for the order of succession. On the other hand however he holds that it is itself no sort of change.[45] Rather, it is presupposed by all changes, in that the latter take place *in* time. For the rest, Kant leaves things as they stood, in particular as regards the unity and infinity of this measure or standard.

With this he arrives at the following further determinations: time cannot be a concept, but must be an intuition; otherwise it would not be individually unitary. Time is, therefore, an infinite intuition. How, indeed, could it be a completed infinity, if it were a concept abstracted from the always finite collection of objects of experience? How could time be realised in them, when it is much rather that in which they are and which underlies them? It is in being tied in intuition to this or that, and to the same or to different parts of time, that the objects of our concrete experience receive their temporal position, their reciprocal temporal ordering, their being at the same time or earlier and later in determinate distances.

Since it is given prior to experience of temporal things, pure time is purely subjective. Strictly speaking there exist no temporal things. If things appear to us as temporal, then this temporal character is due to our subjectivity alone. And thus also Kant calls time the form of inner sense. (Admittedly with little appropriateness, and therefore also for many misleading, since time is not itself meant to be the temporal character of things but only the reason for the temporal character of things; for it does not itself have succession, temporal perseverance or change, but only the things in it. If it were a form in the traditional meaning of the word, then it would be concept rather than intuition.) The fact that time belongs to our subjectivity implies, according to Kant, that the application of synthetic a priori knowledge is justified and thus makes possible a phenomenal natural science.

However little reason Kant had to be satisfied with the theories of time of his predecessors, we have just as little reason to be satisfied with his own. For the Kantian theory is nothing short of monstrous, both regarding the drastic remedy which he administers and regarding the results to which this leads, and it bears remarkable testimony to the modest capacities of our tone-setting professors that they hold this doctrine of time (and likewise that of space) as fully established and as one of the most valuable parts of Kantian philosophy.

Time and space purely for themselves are supposed to be intuitions. It is clear to anyone who can observe at all, that we have no pure intuitions of space and time. What we perceive in the way of spatial and temporal determinations are nothing but αἰσθητὰ κοινά in Aristotle's sense.

Time and space are to be *infinite* intuitions. But we have no infinite intuitions or intuitions of something infinite, just as we have no infinite positive concepts,[46] as Locke correctly saw and as Leibniz, concurring with him in this, vigorously emphasised.

Where Locke and Leibniz conceive time, as measure of changes, as being itself a change, Kant does the opposite. The whole of infinite time exists, naturally not in itself but only phenomenally, and is unchanging. But all phenomenal things are supposed to exist in this unchanging time and to suffer in it their coming into being and passing away and all changes and to stand in it at a distance from each other. But how can something be in something which exists without itself existing? Thus if something exists in a part of time which belongs to the most distant past or future (for the intuition of time consists, surely, of past, present and future), then this part, because — according to Kant — it exists just as much as the present part, must exist in the same way as that which exists in the present. There may thereby exist changes in time, but these could consist only in temporal displacement, by no means in a coming into existence and passing away. That which exists for a shorter time is similar to a less extended thing that exists in space: it is equally actual in all its parts. Every thing persists in its being without beginning and without end, but subject to constant temporal displacement.[47]

It would be uselessly prolix, where such crude violations have been exposed, to continue the criticism still further. Otherwise I could emphasise that it is evident that everything must exist in time, not merely phenomenally but also in itself, and also that what is temporal is grasped with evidence in inner perception. And further still, that even if one could prove the a priori subjectivity of time, it would still be logically unjustified to pick out certain subjective prejudices — which Kant, under the name of the 'synthetic a priori', wanted to take as basis for all science — and make them into the foundation of inquiry into the temporal succession of our phenomena.[48]

Kant asserts further in support of the subjectivity of time and space that the assumption of their existence in themselves would lead to true monsters. Infinitely large things — and these neither substances nor accidents! And he is not wrong. But how does he arrive at the idea that they are monsters? Clearly only in comparison with other things which he accepts as existing. But he in fact accepts nothing as existing other than phenomena. Even as phenomena, then, as merely subjectively considered, they would have to give the impression of monsters. And in truth does not the unitary infinite space appear in all times and the unitary infinite time in all spaces? And how would this be thinkable without contradictions, if time did not undergo a spatial duplication of its parts? When however Kant adds the qualification 'without being substances or accidents', then this appears not to be correct according to the old usage of the term 'substance', for according to this the word signifies nothing other than something which is for itself, rather than as property in something else. Space and time are supposed after all to be *pure* intuitions, and thus to be intuited for themselves. This temporal substance would also have accidents, to the extent that it would be in all parts of space. It would be characterised by a here and a there (and conversely for the spatial substance)[49] and also by infinite spatial extension in three dimensions.

All this shows, as I said, that pure time and pure space, in the way Kant describes them, are monstrous not only as such but already when taken as subjective intuitions. And therefore such a time and space are in truth not present in intuition.

12. *Schopenhauer.* The outrageousness of Kant's 'Transcendental Aesthetic' has not prevented its finding many supporters. Even Schopenhauer valued highly Kant's doctrine of space and time and held on to it, though with some additions and with modifications of which he was perhaps not even clearly aware. Sensations, which Kant held to be possible from the very beginning only under the forms of space and time, are held by Schopenhauer to make their phenomenal appearance initially in and of themselves; only then do they gain, so to speak by association, their spatial and temporal positions, though the rapidity with which this process takes place means that it escapes our attention. He conceives sensations as an empiricist conceives qualities, except that it is not empirical but rather a priori moments that become fused thereto. Further, space is not supposed to attach itself to sensations in all modalities. Those of hearing are not subject to this connection, while on the other hand the connection with time, which is in this case primary, is supposed to be only secondary in the case of space. Kant's whole intention is here misunderstood, and what is impossible becomes so to speak an impossibility of higher order.

Time is according to Kant that in which all change proceeds but is itself something that does not change. But Schopenhauer, apparently without being aware of his opposition to Kant, teaches that it is in a continual ceaseless uniformly proceeding flux. It is not merely that one thing is after another in time, but also that one part of time itself is after another. In this he approaches Locke and Leibniz (though because of the different underlying conceptions he is still very far removed from them). We have seen how Kant, in consequence of his contrary assertions,[50] leaves himself quite peculiarly exposed to attack. But if Schopenhauer avoids the Charybdis, he still falls into the Scylla. If time is to be not unchanging, then it is always another, and then another. Thus there exist not merely other and then other parts of time, but another and then another time as a whole.[51] And thus also many intuitions of time occurring in succession. Or does Schopenhauer believe that something could change without becoming something other? This could indeed be the case, given that he also believes that time should have many parts but that none of them would be different from any other. (This he says explicitly, in spite of the law of contradiction.)

It is very strange also when Schopenhauer, who holds on to the idea that time is an intuition (it is, he says, intuitable), goes on to add: but only as an infinite straight line. This reminds me of a book where it is said that the soul sees the body in a dream, not however as body, but as something quite different. And it is quite peculiarly strange when he says that time is to be grasped as flowing. If the intuition of time presents a line, then it does not present a change.

Regarding the additions which Schopenhauer made to these modified Kantian assertions concerning time, the most essential which I want to mention here is Schopenhauer's identification of time as the principle of individuation. That the consequences of this proposition are not in harmony with the remainder of his assertions was something of which he himself was not clear. Everyone who is even to some degree informed about the problem of universals knows that no universal exists for itself, but can exist in reality only as individuated, i.e. in full identity with individuals. It is moreover indubitable that the question regarding the principle of individuation relates not to an external principle of the thing (as for example efficient force and end), but rather to an internal principle. Just as in the case of the principle of specification it is a matter of a *differentia specifica*, so here it is a matter of a *differentia individualis* of the individual thing from other individual things of the same species. Thus, just as the specific difference exists as many times as there are things that participate in it, so must the principle of individuation exist in the things as many times as things are individuated by it. But there is, according to Schopenhauer, only *one* time, and thus it is impossible that it should be the principle of individual duplication.[52]

Perhaps someone will say here in defence of Schopenhauer that, while time is one, its parts are many, and different individuals are individuated in having assigned to them different parts of time. But this answer is not meet, since after all many things exist at one and the same time, as for example in the present instant. These things can therefore not be differentiated from each other as individuals by time. But even if they were, then the now would belong to each of the individuated things, so that there would be as many nows as there are presently existing things.[53]

Moreover, for everyone with sufficient knowledge in regard to the problem of individuation, it is clear that, if time is to be the principle of individuation for the accidents, then it must be the substance of the accidents (which already Aristotle sufficiently stressed).

13. *Herbart*. Thus neither through modification nor through additions to the Kantian doctrine of time has Schopenhauer arrived at anything of tenable value. And something similar must confront everyone who has not freed himself completely from the basis of the Kantian conception of time. Thus Herbart, in particular, has done nothing to protect the phenomenal (subjective) time which he takes over from Kant from the charge of inconsistency. The 'intelligible' time, which he takes as real, shows phenomenal time transformed to such a degree that it appears as sheer equivocation when he refers to it, too, as 'time'. In its case, he holds, there obtains no contradiction. In the case of that which exists merely phenomenally, however, he holds it to be perfectly possible that it should contradict itself. This is however a great and for him consequential error. That which is contradictory is certainly possible in concepts, in unitary intuitions however it is just as little possible as in reality.

14. *Lotze*. Lotze, much more than Schopenhauer and Herbart, has distanced himself from Kant in his conception of time. Thus he asserts that we possess no intuition of time at all, and that time is not merely phenomenal but real in itself. The Kantian antinomies in relation to time make no impression on him. If they counted against any mode of actuality of time, so also against a phenomenal actuality. 'Past', for Lotze, means so much as 'one-sidedly conditioning what is present', 'future' so much as 'conditioned one-sidedly by what is present'. The most immediate past is that which conditions the present immediately, the more remotely past that which determines the present mediately. The most immediate and the more remote future are distinguished similarly, so that we arrive at the idea of an infinite time, as also (when reflecting on a merely possible continuation on the side of the conditions) at that of an empty time.

That which is past and future exists only in our mind, that which is present however is real also in itself. The succession of causation is the most intrinsic nature of that which is real, since that which is real is not allowed to be plucked apart into essence and existence, content and reality. Thus according to Lotze time, or strictly speaking the now, is the nature and the being of things and events themselves. Lotze admits that we feel ourselves in a puzzling way under pressure to include also real becoming within the compass of an enduring reality, but only the philosophy of religion with its reference to the divinity can yield an answer to this puzzle.

In his *Grundzüge der Psychologie*, too, he writes: No sort of effect of one element on another is in fact thinkable without contradiction unless we consider them all merely as dependent modifications, set in constant mutual relation to each other, of one single truly existing being, which is in them all the ground of their existence and the ground for whose sake they must exercise a determinate effect in determinate circumstances and the ground for the fact that they can exercise this effect.

The change of scene is considerable! Lotze's conception has practically nothing in common with those of Kant, and with that of Schopenhauer only perhaps the rejection of the Kantian unchangeability of time. It is as if we have been restored to medieval times, when one spoke of an *internal* time of every natural thing, defined by Bonaventura as the succession in the existence of created things. To be sure, Lotze would not have approved of his separation of existence and essence; he would rather have followed Suarez and others in identifying the existence and nature of that which is. But he would thereby have distanced himself from Bonaventura only so far as to extend to all things what he had accepted for only a part of things (namely a succession in their *nature*).[54] We find also in Bonaventura that time is brought into relation to the production of effects, namely — and Lotze, at least in the last instance, also casts his eye in this direction — to those of the God of creation.

Something else that reminds us of the scholastics is the fact that Lotze believes himself free in his metaphysics to dispense with painstaking psychological investigations. Thus he sees himself only rarely called upon to incorporate psychological observations (for example that intuition and presentation are of essentially different character; that it would be contradictory to suppose that a change could be grasped in an intuition, since it would otherwise have to represent a certain stage of the change, which is a matter of presentation, etc.). This lack of psychological depth contributed not a little to the failure of his attempt. What could be more inappropriate than mixing up the concept of effect with the general concept of time and of speaking as though all that is immediately preceding is already for this reason a condition for all that is immediately following? If, in the instant the balls here in the bowling alley reach the skittles, a prince of China is born, then this is clearly a case of before and after, but it is certainly not, or at least it is not immediately clear that it is, a case of conditioning and conditioned, even if speculative profundities in the philosophy of religion would perhaps want to make this believable to us by appeal to the one unitary essence of all things. There are today positivists

66

enough who call into question the very concept of causality while holding on to the fact of temporal succession. And who could assert that they would contradict themselves thereby?

This precariousness of psychological foundations makes itself felt over and over again in Lotze's writings. If intuition grasps only what is effected, presentation only what produces effects, but if active conditioning and passive being-conditioned are yet incapable of being grasped each for itself, then these can be given to us neither intuitively nor in presentation. And for this reason also one cannot see how, since that which is immediately past is presented by he who remembers and that which is immediately future by he who expects, these two are to be distinguished from each other in their content.[55]

I am in agreement with Lotze that the things in themselves are temporal. But because he has surrendered all evident intuition of a before and an after, his demonstration of the reality of time is obtained by pure deceit.

I do not want to repeat objections already raised against scholastic doctrines. I remark only that a production of effects is like a motion in that it can without contradiction proceed either more slowly or more quickly, while the passing of time must be regular in all its parts.

Thus Lotze's attempt, too, has failed, and thereby also certainly the last serious attempt to afford for us a deeper insight into the nature of time.

15. *Hegel.* I have not included in this series what Hegel has to say about time, since his entire philosophy is today no longer able to be taken seriously by anyone who is himself to be taken seriously. But still, since Windelband is in this respect of a different opinion, it will suffice to satisfy him (and to amuse the reader) if we quote one or two of the pearls of Hegel's wisdom: 'The negativity, which relates to space as point and develops in it its determinations as line and surface, is however equally for itself in the sphere of being outside itself..., thereby however appearing indifferent against the still side-by-sideness. Posited for itself in this way, it is time' (*System der Philosophie, Part II*, §257). It is 'the negative unity of being outside itself.' 'It is the being which, in that it is, is not, and in that it is not, is.' He refers to it as something 'absolutely abstract, ideal' and in the same breath as 'intuited becoming' (§258). It appears superfluous to waste here any word of criticism. Assertions such as this find death by their own hand.

16. *Bolzano.* It is rather Bolzano, among the thinkers before

Lotze who concerned themselves with the clarification of time, who would deserve to be mentioned. Bolzano was little considered in his own day, but more recently he has been granted special notice by those mathematicians who have approached the borderlines of philosophy. In regard to space and time he propounds a quite peculiar doctrine.

He asserts of both in his *Paradoxes of the Infinite*[56] that they are in themselves, but that they do not exist,[57] are not realities. They are not substances, and not qualities of the substances, but rather determinations of them. In regard to time, in particular, he declares that it is nothing changeable and that it is that determination to be found in every changing (that is, dependent) substance the presentation of which must be added to the presentation of this substance if we are to be able, in regard to two contradictory qualities b and not-b, truthfully to ascribe to it the one and deny the other.

That neither past nor future time would have existence creates for him no difficulty in conceiving time as an infinite continuum — for after all, he says, even the present has no existence. But from this it does not follow that time is *nothing*. Propositions and truths in themselves are *something*, too, though it will occur to no one to hold them to be something existing.

Bolzano wants to derive from this doctrine of time also the doctrines of the science of space and for example show that (and why) space has three dimensions, etc. He conceives space as the sum total of all places in which the created substances are located. The places however he conceives as those determinations of substances which supply the reason why, for a given stock of qualities in a given time, they bring forth in each other precisely these and these changes.

Many will initially be in the dark as to what he is to understand by a determination that is itself neither substantial nor a quality of substances, but is yet in the substance. This can perhaps be sufficiently elucidated through examples like 'thought', 'loved', etc. A real thing which I think possesses the determination that it is thought, one that I love that it is loved. But one will not call this a real property. And just as little does something have a real property as a result of the fact that it has the determination of being owned by me or of being married. If, however, all real determinations of a thing remain the same, then such non-real determinations cannot now apply and now fail to apply. Yet Bolzano affirms just this of the spatial and temporal determinations, and I must therefore dispute that his assertions about space and time are thinkable without contradiction. One says that in order for a thing to acquire or lose such a non-real determination, some change must really occur, if not in it then in some other thing. Something cannot begin to be thought unless some thinker begins to think it. Now I ask where, in the case of changes of temporal determination, is this other? Bolzano would have to show us this thing, if he is to clarify the nature of time. He does not and could not do so, for it is evident that a thing would still change its time even if nothing existed outside it.[58] Thus this temporal determination appears to be something real in the thing, or at least something that is not able to be different when everything real in it remains the same.

In regard to Bolzano's attempt to derive the three dimensions of space from the nature of time on the basis of such a conception of space and time, we can be sure from the very start that it can just as little succeed as all the attempts that were made by Leibniz, Kant, Hegel, Lotze and others to demonstrate the impossibility of topoids of more than three dimensions.

17. *Wundt*. Among philosophers since Lotze and who are still our contemporaries it may well be Wundt, in particular, whose opinion will be of interest. Is it not he after all who has, with an almost superhuman productivity, flooded the philosophical market with his books as the clouds of the flood poured forth their waters? And Wundt has of course also allowed his voice to be heard in regard to the nature of time. His ideas on this matter are very original. It appears to him to be not impossible that time, which one normally conceives as proceeding in a straight direction, could have a slight curvature. However, those who do not want to bring their healthy common sense as a sacrifice to the altar of this authority will reject as nonsense the suggestion that the same point in time could lie both before and after another. In any case it shows that its author has strayed completely from the true concept of time. Yet Wundt is not aware of this, and so seriously does he take this possibility that he believes himself able with its aid to escape the conclusion of the universal heat-death to which the law of entropy which he accepts threatens to lead. Yet it is manifest that he is entirely unclear as to the way in which this escape is to take place. One may ponder this argument: if no heat-death is to occur, then the law of entropy leads to ever further decrease in warmth,[59] say from moment A, whether this lies in the present or in the future. Since, however, contradictory determinations cannot be at the same time and thus also not in the same moment A, it must follow,

contrary to the law of entropy and if we are not to violate the law of contradiction,[60] that the quantity of heat that now exists will reconstitute itself with the new occurrence of the moment A.

I called this idea original, and yet here, too, it still holds good that there is nothing new under the sun. For was not Suarez already of the opinion that God could allow individually the same time to return again, so that it is almost as if Wundt on this occasion appears as a sort of neo-scholastic, whom he otherwise describes (and not in order to recommend them) as kith and kin of, among others, Locke and even myself (though in fact he possesses sufficient familiarity neither with our nor with the scholastic doctrines to be justified in making such comparisons).

18. *Renouvier.* This, then, is how things stand concerning the most loudly praised of contemporary German philosophers in relation to the question of time. If we cast our eyes in the direction of France, then we find the respected head of a school proclaiming a peculiar doctrine as to time. Renouvier, a monadologist influenced in much of what he says by Leibniz, departs from him in that, while conceding continuity to phenomenal time and phenomenal space, he holds real time and real space, i.e. that which he conceives of as existing in themselves, to be discrete. An hour, as he conceives things, is made up of a finite number of temporal points. He was led to this view by synechological difficulties. It escaped his notice, just as it did Herbart's, that what is contradictory cannot only not be given in the world which is in itself, it cannot be given in intuition, either. But that a time cannot be made up of a finite number of discrete points is clear already from the requirement that cause and effect have to be in contact with each other temporally as before and after, which would be impossible given the discreteness of time. Moreover, every thought of a true before and after in separation from the thought of a real or possible continuity, is a *contradictio in terminis*. Thus under such conditions I do not even need to go into the physical unwholesomenesses that would flow, from the assumption of a fixed number of points in a length of time and space, for the larger and smaller velocity of proceeding motions. Renouvier takes refuge here in his incorporeal monadology. It could however be shown how he is constrained by such refuge to abandon the whole rich treasure of the natural sciences.

II. On memory

Dictated 23 April 1913 *[Ps 10]*

1. The multiplicity of what one calls memory. When one deals with memory there is a plurality of occurrences which must be kept apart:

> i. something is sensed as past,
> ii. something recurs as a matter of habit,
> iii. one has instinctively the correct belief that one has experienced something earlier.

i. This is the case where we, as we say, *hear* a speech or melody, *see* or *feel* a movement or rest and in general *sense* any kind of change or perseverance. Here phenomena appear to us in a certain temporal succession, just as certain other things may appear to us in spatial succession. However the differences in the former case are given through modal differences of the presenting (and of the acknowledging based thereon), in the latter case through differences in the objects presented.[61]

In the case of memory I do not appear to myself as one who has earlier experienced something as present, but rather as one who now experiences something as past.

ii. An essentially different case of memory is involved when I am in a position to reproduce in myself certain of my earlier thoughts, as this habitually occurs according to the laws of association. Thus one repeats a poem learned by heart, perhaps without even thinking that one has heard it before. I am often uncertain, in the case of a melody which occurs to me, whether it is original or not. Only when it is not do I have a case of memory, and this is so even if it is one that is not characterised as such. I then appear to myself as one who has experienced something in the past just as little as in the earlier case.

iii. In contrast to this, however, are other cases of reproduction where I have the belief, indeed the full conviction, that I have to do with the reproduction of something experienced earlier.

There are still many differences which may occur here. Thus sometimes the belief is bound up with a certain doubt (one says: I think that this is known to me already), and sometimes it is not. And again, it sometimes remains uncertain to me in my belief whether the time in which I am supposed already to have experienced something lies near or far, indeed whether something has appeared to me once or several times, and if so how often. Sometimes however, as for example in the case of a refrain, or in the case of the similarity of a melody with an earlier passage in the same piece, my belief ascribes to the earlier appearance a practically definite distance in time.

2. Greater and lesser perfection of memory. Just as we can speak of a greater or lesser imperfection in this connection, so also in regard to the exactness and completeness of what, as one says, comes back to me in memory. He who has heard my speech may perhaps be able to repeat its main ideas, but he could not recite it verbatim.

3. It is not necessary for memory that one relives the event in a less vivid way. One must not believe that it is necessary, in order to remember something, that one relive it again in a less vivid way or with less vivacity. He who regrets an earlier desire has a memory of this desire, he is however far from renewing in a present desire that which he mourns having wanted. And just as little is it the case that I fall back into an error when, in order to recant it, I recall it to the memory of myself and others: I do not after all share my opponent's belief when I am attacking it.

4. The psychic events which I recall in memory appear not as secondary but as primary object. It is impossible that my belief that I had earlier desired and judged in such a way should be a relation to something as a secondary object. We must rather speak here of a relation to something as primary object, even though this is here something psychical. The relation to a secondary object occurs always in the *modus praesens*; the *modus praeteritus* applies exclusively to the relation to the primary object.

5. When I sense something as passed, this is not a matter of inner sensation. Even if I sense a tone as past, it is precisely a past tone that is the object of my sensation, not a past hearing.

6. That which is presented in full definiteness[62] as present must be presented as beginning or ending or as continuing to exist. The present can exist only as boundary of the past or of the future or of both, and if something that is present is presented with full determinateness of temporal *modus* then it appears indubitable that there must be included in the presentation its being given as boundary of the past or of the future or of both. Yet still, it is difficult to grasp how such a fine differentiation should take place if not even the smallest part of the past or future should be

co-presented, so that their observation would be quite simply impossible. In fact one must in general distinguish between inner observation and inner perception.

But just as every line segment is demonstrably different for different plerosis and different teleosis, so, too, is every point. And these differences are such as to apply to every single point. Otherwise one motion could not lead to a more distant point than another in the same time, in that, with a quicker or slower passage through a given stretch, the quantity of local determinations would decrease and increase with time. Also, the lengths of the circumferences of two concentric circles could not be different.

When an inference is made, the thought of the premise produces the thought of the conclusion and this is noticed by us.[63] This implies however that we notice how the two are in contact with each other temporally, for that which produces the effect and that which is effected meet together in time as end and beginning. Here however the temporal contact and the effecting of the one through the other is itself such as to take a certain time (which is to be explained by reference to the example of a body which pushes another body before it while also bordering on it continuously in space).

7. In what sense is the *modus* of the present not determined? Its exact[64] determination does not require the specified and individuated presentation of the object that is thought as present.

8. The question as to what is the standard for the estimation of temporal distances in the third case of memory. However easy it is to describe the relevant temporal differences when a succession is *experienced* in sensation, it is just as difficult to give an account of how these differences reveal themselves in other cases.[65] It is after all a no less complicated matter to specify what it is that serves as reference-marker for our determination or estimation of spatial distances in what is seen. Here, as in the former case, the most diverse moments may work together.

In other respects, too, analogous questions arise in regard to both space and time. Thus in relation to the visual sense one has asked whether in the intuition itself the local determinations always reveal one and the same surface, so that it is only in our judgment about the real external relationships that there arise different estimations of depth — and indeed also (as for example in the Zöllner figure) of length and breadth. Similarly, one could ask whether our intuition of time manifests always one and the same temporal stretch receding from the present so that it would only be our judgment about the temporal distance of actual historical events which would vary — to the extent, indeed, that in the determination of the before and after we are induced to think also of something that is *later* than the present and to expect it as future. (Perhaps the latter conception is the correct one.)

9. Are the phenomena of memory of the third case repeated in intuition in the *modus praesens*, and are they declared to belong to the past only through judgments? Does the sensory field of time allow itself to be intuitively changed and extended just as little as the spatial field of vision? If it is right that we are capable in relation neither to the past nor to the future of enriching our original stock of temporal *modi* by others, but are able only to think up relations of distance arbitrarily by analogy with what is given, then from the fact that only external perception manifests a continuum of temporal *modi* while internal perception shows what it shows always and only with the *modus praesens*, there follows the impossibility of our intuitively presenting ourselves and our psychic activities other than with the *modus praesens*.[66]

Someone might perhaps believe that this does not harmonise very well with the fact that we speak of past times in our lives and that we ascribe to one event recalled in memory an earlier time, to another event a later. Yet the fact that the mode of intuition is universally the same is compatible just as much with the greatest differences of temporal estimation of events as is a difference within relatively narrow limits. A comparison may clarify this matter. We said already that we grasp our own substance not according to any sort of substantial differences but only acording to the broad and quite general concept of substance. Thus, as to our own substance, we appear to ourselves as not at all different to any other human soul. And if we have reason to believe that there are accidental differences, for example to believe that another judges or wills differently than we do, still, it is possible without contradiction to conceive that there has been brought about an equality in all of these respects. Yet we should still acknowledge the substances as different substances. In the same way also we regard two events in our life, notwithstanding the fact that we present them with the same temporal mode of intuition, as belonging to different times, and we do this even if we think ourselves in both cases as carrying out the same psychic activity. Indeed we can, according to the indications at our disposal, ascribe in our judgments widely separate positions in time to that which manifests itself intuitively as entirely the same.

10. Guyau's empirical theory of time. The view of Guyau is interesting. Guyau conceives spatial distances as being employed in the assessment of temporal relations, and he draws attention to (among other things) the fact that one says of the future that it lies before us; of the past, in contrast, that it lies behind us. But in embracing in regard to time an extreme empiricism which does not even acknowledge that intuitive temporal differences are present when I hear a melody or grasp a spoken word, he is unmistakably in error. Space can be of service in our temporal judgments only in a way similar to the way in which written signs are of service when we substitute them for spoken signs. The services rendered by such signs in calculation are familiar enough.

11. The belief that the events revealed to us in memory belong to one and the same individual life. Among the different cases of memory we have distinguished, there is one in which we are of the belief that we have experienced something earlier. Everyone thinks himself thereby as a unitary individual, as a unitary individual substance identical with that of the thinker of this substance. Now this would give rise to no problems at all if each one of us were able to grasp in inner perception his own individual substantial difference, so that in memory, too, he would intuit himself as determined by this individual substantial difference. But this is not at all the case. The difficulty can however be resolved without much trouble even under the conditions which in fact prevail. If the self sees himself in memory only in his most general substantial determinations, then this is analogous to the case in which he grasps himself in inner perception merely in his most general substantial determinations. Now this does not stand in the way of his coming to know that the substance he apprehends is an individual and single substance identical to that of the one who apprehends it. But then similarly in the case of memory, also, grounds are not lacking for the belief that one has to do here with one single individual who is identical with the substance of him who has the belief that this is a historical substance which has a certain accident it experiences. A whole series of such memory phenomena assembles itself together as belonging to the course of my own single life. Admittedly, one has here no evidence either for the veracity of the individual facts or (and still less) for the identity of all these substances which appear to me in it. Yet still, if we become aware of the fact that we are here led by a blind impulse, then the hypothesis which naturally suggests itself, like the hypothesis of an external world of bodies, has already proved itself so often to be

reliable that I cannot find it unreasonable to continue to believe therein.

Addendum

1. Since the temporal differences that are experienced in sensation relate only to the primary object, it is easy to see what happens when one has sensations which go on for some time. The small temporal field of external perception remains always the same and is merely filled now in this way, now in that. Inner perception however, which manifests always the one *modus* of the present, maintains this for the different events following each other. This has led some to the assertion that time remains always one and the same and only the things change in the same time. But what does this mean, 'they change in the same time'? Surely this: that before and after are different things in the same time. But how can a before and an after be in the same time? Does this not mean assuming of time that it would itself permanently exist within another time?[67]

> You say time passes
> Because you do not understand,
> It's time that stays
> And you that passes

— an obvious absurdity to which one has been led through the confusion of time itself with a lasting peculiarity of our sensory intuition, which presents things always with the same temporal *modus*! It is true also that something is always present and that something has existed and will exist in every moment of time, but not the same something.

2. The question arises whether something can maintain itself completely unchanged in every respect. One would then have to acknowledge it at the same time with a variety of temporal modes, whose magnitude would correspond to the distance from the first of these modes to the last. Its perseverance would not be a course which proceeds through time. But would it be a matter of its being at rest? Or does rest itself require a change in some respect while everything else would persevere throughout this change? Would it otherwise be possible to distinguish a greater or smaller amount of

rest? It can be shown that even in God himself, the first cause of all things, there obtains a continual infinitesimal real change and that this must have as consequence a continual real change in all that is conditioned by God. One must guard against confusing these real differences with the equally regularly changing temporal *modi* of correct acceptance of this very fact.

3. One often does not bring to clear awareness the difference between indications and that which they are indications of. Thus for example one holds the localisation of the hand to be phenomenally different when one leaves it hanging or lifts it up, brings it to the mouth, etc. It remains in all of these cases the same, but the relation of the actual position of the hand to that of the remainder of one's body is differently judged. This confusion played a role also in the dispute about the dimension of visual depth. Some hold that this dimension appears only with binocular vision and that for the same nerve-ending it is subject to considerable change in virtue of various sorts of combined action of the eyes. Think of the different impression stereoscopic images of paintings make when compared with perspective drawing and colouring even under conditions of monocular vision. But still this doctrine, which leads to unspeakable complications, is in my opinion not acceptable; much rather do we have to do here with a case of the most powerful illusion of judgment. Think of the way that judgment gets caught out by the image of a staircase going from top to bottom or by a hollow relief seen with *one* eye.

Thus it is understandable that in regard to time, too, one often treats as intuitive, differences that are in fact only to be conceived as differences of judgment on the basis of indications, the peculiarity of which one has not at all made clear to oneself but which yet manifest a most powerful determining influence.

III. Our intuition of time is a continuum of modes of presentation and acceptance

Dictated 2 November 1914 *[T 14]*

The text that follows is a version of the original dictation abbreviated by A. Kastil. [Editors' Note]

1. That which is continuous can be classified from a number of points of view. One of the most important is that according to which it is partitioned into what exists as a whole and what exists only in a boundary. A continuum of the first class is for example a body as spatially extended. A continuum of the second class, however, is everything that proceeds in time as such. We wish to refer to the one as *spatially*, to the other as *temporally* continuous. Each thing, in so far as it endures and remains for a period of time entirely unchanged in some respect or is subject to continuous change, appears as something temporally continuous. Thus also every body is something temporally continuous, not indeed in regard to the three dimensions in which it exists in its entirety in every moment, but in so far as its existence temporally continues, whether with some sort of change or as quite unchanging.[68]

It is not only bodies and the surfaces and lines that can be distinguished in them that are spatially continuous, and not only those qualities which they possess in some extension and perhaps other properties — but also topoids of other numbers of dimensions (as these, in the general opinion of mathematicians, are conceivable without contradiction). And then of course also their boundaries, which manifest an extension of their own. Further, our intuitions not merely of what is spatially extended but also of what is proceeding in time in so far as this is present to us as a whole.[69] A motion from place to place, although in a certain sense to be called spatial, is for all that still — because it proceeds in time and because it exists not in its whole extent but only in one of its boundaries — to be counted among the temporally and not among the spatially continuous.

2. I want now to pay closer attention to the peculiarities of what is temporally continuous and of our presentations thereof.

It is clear that every movement exists not as a whole but only in one of its boundaries. It seems that Aristotle wanted to give expression to this truth when he referred to movement as an imperfect reality and paradoxically spoke of it as the reality of the possible as possible or of the reality of that which is in possibility as such.[70] This is, as he remarks, difficult to grasp, but still not absurd. Indeed if a movement is in the middle of its course then it exists neither in its first half nor in its second half but only in that moment in which the two halves are in contact with each other and which establishes the connection between them. Yet this character- isation is very inadequate. We would not be able to say that a movement proceeded from a to c if it were correct only that it existed in a certain moment — for example in the moment where it passed point b — and in no others. We must also be able to say that it *has been* in every one of the points between a and b, in the one a longer, in the other a shorter time ago, and that it *will be* in every one of the points between b and c, in the one in a nearer, in the other in a more distant future. There lies in this a certain acceptance of factual occurrences, but not as existent or present, and this shows that there is a type of acceptance which does not accept as existing yet still distinguishes everything historical and everything that is to come from that which never is, nor was, nor shall be. When that which was first of all given as present appears as more and more past it is *not other objects* which are accepted as existing, but the same objects which are accepted *in a different way*, with a different *modus* of acceptance.

This *modus*[71] is however not something unitary: in reflection of the continuity of temporal intuition it is a *continuum* of *modi*, first of all of presentation and then also of acceptance. That which is past is after all not presented as equally past, but as more and less past.

3. There is however no shortage of attempts to render superfluous the assumption of this modal continuum. Thus I hear assertions which suggest that acceptance, as this applies to what is historical, is not an acceptance at all, but is rather to be called a rejection, or that one ought to speak here of a third quality of judgment.

The latter would however amount to the giving up of the law of excluded middle.

If however, in the case of what is historical and of what is in the future, one were to speak of rejection instead of acceptance, then the foundation for the relation of the earlier to the later would be thereby done away with, so that the relation itself could no longer be granted any truth. Yet we do correctly say that we are 100 years later than those who were 100 years ago, and that they were 100 years earlier than we are. Not only this, but also that everything that was 100 years ago was earlier than something that was 10 years ago, and that everything that has been has been earlier than something that is to be.[72]

4. Another attempt to manage without such a modal continuum seeks to put in its place a continuum of differences in the object. One admits that the past and future parts of a movement are not to be accepted with the same *modus* as its currently existing moment, but insists that this divergent *modus* is something *unitary*, common to all that is past and future. The differences of past and future and of nearer and remoter past and nearer and remoter future are entirely differences in the objects. An object appears as a different object according to its distance from the present in the one direction or in the other. Like every different position in space, so also every different position in time would be something that somehow affected and determined the object. And thus for example all things existing in the instant of the birth of Christ would have a common specific determination as things to be found within this instant. And similarly those falling in the same instant one century later.

Yet whatever truth there might be in this, under no circumstances do the differences of which one here speaks coincide with those of future, present, past, more or less remote future, more or less remote past and so on. For the instant of the birth of Christ, and thus also every other instant, belongs successively now to the future, now to the present, and now falls in the class of what is past. If a tone in a melody appears to us first as present and then as belonging to the most recent past, then we do not believe that it has two temporal positions, or that it has exchanged the one for the other, but rather only that it is for us, who have continued to exist in time, to be accepted in a different way. From something that is present it has become for us something that is more and more remotely historical.

5. The assumption of modal differences is therefore not to be avoided. And in addition to the *modus* of the present a unitary temporal *modus* for everything past and future would by no means be sufficient. For a distinction in ways of acceptance similar to that which arises between what is as opposed to what was or will be arises also as between what was earlier or later and what will be earlier or later.

a) Already language points in this direction when it distinguishes, beside the perfect also a pluperfect tense, and beside the future also a *futurum exactum*. It is, admittedly, true that everything past and everything future shares in common that it is *not*, and that its mode of acceptance is not the *modus praesens*. But this establishes just as little against the view that there are within the past and future further differences of mode of acceptance as the circumstance that curved lines have in common that they are not straight establishes that their mode of curvature is not different according to whether they are strongly or less strongly curved and according to whether they are curved on this or that side.

b) Indeed, the presentation of a movement, as contrasted with that of a spatial stretch, is not merely impossible without two modes of presentation, the one presenting as present, the other not; it requires a further continuity of specifically different modes of presentation. If, however, I were to present the whole motion as past (which I can of course do), then there would no longer be any talk of a difference of modes of presentation and perhaps of acceptance.

c) Perhaps the following will clarify further what has been said. It is sometimes held that a single spatial point could not exist if there did not exist a body to which it would belong. Given, however, that there exists in truth only what is present, this is not correct. Consider, for example, the apex of a pyramid that is destroyed gradually by God from the base upwards, so that the topmost point would exist alone in the very moment in which the destruction is completed. It is sufficient if, as in the present case, a body to which it belongs has previously existed or will come to exist, for God could also allow the pyramid to come gradually into existence, beginning with the topmost point alone. Suppose, however, there were no difference in mode of acceptance as between something set further apart in time from the present moment and something lying in immediate temporal proximity thereto. A single present spatial point which existed alone might then just as well be the end point of a pyramid that had existed one year ago and had been gradually annihilated as of a pyramid which existed immediately before it. But this is not the case, and the reason for this is connected with the fact that every boundary is determined in its nature by that which it bounds: it is related thereto as inner boundary and not merely as a *terminus extrinsecus*. And such determination is brought about not merely by what exists simultaneously but also, as our example shows, by what is earlier and later. But this earlier and later must have been or be going to be in immediate temporal proximity. How could this be compatible with the idea that the mode of acceptance should be entirely the same in the case of something that was shortly before the present and something that was 1000 years ago?[73]

d) The theory which assumes only one single modus of the past is subject to the objection that it virtually transforms the presentation of a temporal passage into that of a topoid of one dimension. Certainly it still allows one to talk of before and after, but it transforms this before and after into an analogue of spatial juxtaposition as this would pertain to every topoid, whatever its number of dimensions. Only someone capable of a presenting with different modes and of a continuously changing mode of presentation can have a presentation of rest and motion, of continuing to exist or of proceeding in time.[74]

The analogy between time and space should not, then, be allowed to be overestimated. Thus a chronoid of more than one dimension is directly absurd, and one could just as little conceive of something temporal which would show itself as real and individual through its time-determination alone. It would appear entirely without objectual content,[75] where what is spatially extended was as such identified by Descartes with the substance of bodies.

6. The conclusion is that the differences of present and of nearer and remoter past do not appear as differences of temporal objects analogous to differences of location. But this should not be taken to rule out the real existence of true temporal analogues of such local differences. Nor should it be taken to refute the thesis that continued existence in time, even given the most conceivably perfect sameness of that which endures, is never thinkable without a continuous material transformation. This becomes clear when one considers that every period of rest has a certain length and that this, given that it is a magnitude, must allow the distinguishing of a plurality of parts. Yet there is no plurality without difference. Thus also there is no plurality of succession without variation.

One can approach the matter also in the following way. It is certain that there exist changes in the world. These changes may be caused by something that is not given to us directly in experience. And however this something may be to be specified as transcendent, one thing is certain: that it must itself be subject to some sort of change; for where rest may perhaps come out of motion, it is impossible that motion should come out of rest. Thus something completely changeless could never ever become the principle for something changing. Thus a continual real change is manifested even by the immediately necessary principle itself which then impresses a correspondingly changing characteristic trait on all that it conditions. And thus everything that exists together, that is, everything that exists in the present and so also everything that is in any given degree past or future, must share a common real trait, even though this is not available to us in intuition. One could say that if it were intuitively available then the first principle of all things would have to be copresent in this intuition also. We should therefore, if this first principle is a divine intellect, have an intuition of the same. (Of course one customarily holds it as having been settled that God is without any change, since such a change would signify a transition from less good or to less good. Such assertions are however very superficial and do not take account of the fact that he who knows all truth must certainly, in order to do this continually, be subject to a change: for not all truths are eternal. One could, in light of the fact that the divine life is itself a continual variation, apply to the divinity in the most exalted sense the name of time. The old mythology which speaks of Chronos as primeval being and first divine principle seems thereby to be somehow in contact with the truth.)

7. There are some who speak of *the* time, as of an infinite unitary absurdity, an unreal something that would in some peculiar manner condition things from the outside.[76] They consider what exists simultaneously not merely as existing in *equal* times but as existing in *one* time. Already language, they claim, lends support to this idea, in that one says that things are at the *same* time. For us the place of this absurdity 'time' is taken by the all-sustaining God. His conditioning is his production of effects, and his effects are just so many as the effected things. But now the divine activity of God, as it exists in some given moment of the divine life, lends all that is caused by it a peculiar character trait of the same type. Hence things which exist together are for us not such that they are in just one time, but such that the time of each is equal. It would be quite inappropriate to see in this sameness of time of everything that exists something like a pre-established harmony, for indeed it cannot even be thought without contradiction that the things which are produced in the same moment of the divine life should enjoy the character trait of something which would have been produced in some other moment of the divine life.

IV. Shortcomings of the assumption of a single preterite mode

Dictated 15 November 1914 *[T 16]*

1. That which is spatial and that which is assembled in a way comparable to spatial things,[77] share in common with time and with that which is assembled in a way comparable to temporal things that they appear as continuous. In the former case however we have to do with continua that exist as a whole, in the latter with continua that exist only in a boundary. For the rest, they do not exist but rather have existed, or will do so, and even this never in such a way that they would have existed or will exist other than in a boundary. They exist as a continuous sequence of that which has existed, that which has existed earlier, that which has existed still earlier, and so on, or of that which is to come, that which is to come later, that which is to come still later, and so on. Not only is it the case that two of their boundaries never *are* together, two such boundaries never *were* and never *will be* together.

This allows us to see that in the case of what is temporally continuous we have to do not merely with two modes of acceptance (acceptance as actual and acceptance as non-actual), but in fact with a continuous change in modes of acceptance, and indeed in such a way that the mode of acceptance of something as lying twice as far in the past differs in the same degree from the mode of acceptance as half as far as the latter differs from the mode of acceptance as present. The combination of a here and a there in a past or future is just as much excluded for a body by the law of contradiction as is their combination in the present. Only with *different* modes of acceptance can the two be ascribed to a body without contradiction.[78]

2. When we compare the course of a movement with a state of rest in the same time, then it is clear that there is manifested by the relevant object in the one case more change and in the other less. Differences of place do not enter in the latter case; rather, the local determinations remain completely unchanged. In both cases however the temporal modes of acceptance and presentation manifest a quite regular variation. If we assume that, in the one as in the other, there is given in reality another variation, consisting in modification of the object, then still it would not be excluded that

84

this did not come to appearance with the others, and then in the case of motion in space we should become aware of a variation both in the object and in the temporal *modus*, in the case of rest however only in the latter. And everything speaks in favour of the idea that this is how things are. This is why the layman quite understandably believes in the possibility of something's being preserved completely unchanged in time. If however he conceives time itself as something that is uniformly changing, then he makes of time not a mark of the thing that exists in time (otherwise he would have to assume as many times as there are things); rather he conceives it as a special object in its own right, which, caught up in its own continual and perfectly uniform change, serves us as the measure of the before and after of every change, both the rapid and the not so rapid.

It is in this way that time came to be defined by philosophers of antiquity as the movement of the uppermost celestial firmament in so far as this supplies for us the measure of before and after for all other movements. This shows clearly that these philosophers were able to find no sort of change in the simple perseverance of an object in and for itself. Indeed, there were not a few who refused to say of that which not merely does not change but does not even have the capacity to change that it is temporal at all. It is now however demonstrable that there is nothing at all of this sort. Indeed, even the divine life, to which one has wanted above all else to ascribe unchangeability, is itself in a process of necessary, eternal and completely uniform change, in so far as God knows and wills the course of historical events not merely in every other respect but also in regard to what is present or already past or still to come in the future. In this eternal variation of God, now, there lies the cause of the fact that all created being must be subject to a corresponding change, however much it may continue to exist unchanged in other respects. The fact that it is maintained by God in a different moment of his divine life gives it, moment for moment, a different character. And it is clear, too, that this character must at any given stage be the same for all that is then present. Yet however certain it is that this character can be inferred, it is certain also that it does not appear in our intuitions. It could do this only if we could at the same time behold God himself. And thus also the assertion is again justified that the temporal differences we perceive are only differences of modes of presentation, acceptance — and will.

3. For, if there are some who hold that one needs to distinguish only between a mode of acceptance for that which is present and

another for that which is non-present, and that one could not speak of a continuously varying mode of acceptance in relation either to the past or to the future, then it is perhaps of use to draw attention to the fact that willing (and feeling), too, varies temporally according to its mode. It is something different to want something *de praesenti* and to intend something in the future, and if this difference is clear, then so also is that between an intending for the immediate future and an intending for a later future.[79]

V. The real is the temporally continuous.
There is no internal proteraesthesis.

Dictated 1914 *[T 50]*

1. In earlier investigations I have shown that the differences of present, past and future and the numberless differences within the past and the future are not differences of the object but rather different modes in which we present something, accept or reject something, love or hate something. Our judging and feeling activity is temporally modified because it rests on the activity of presenting.

2. I have for a long time also propounded the thesis that everything that is, is subject in its being and in its properties to a certain uniform real transformation. This underlies all cases of perseverance and change, just as the variation of spatial determinations underlies a uniform extension or a variety of colours or tones in space.

I want now to examine this thesis a little further and will thereby recall above all the demonstration that, as certain as it is that there exists any change at all, so also must the same be assumed of God, both in regard to his knowledge as also in regard to the activity of his will. If something changes, it follows that not all truths are eternal. God knows all truths, and thus also those which are only today. These, however, he could not have known yesterday, since then it was not they, but certain others, which obtained. Thus he now knows for example that I am writing down these thoughts. Yesterday, however, he did not know this, but rather that I will write them down later. And similarly he will know tomorrow that I have written them down. Everything is in perfect accord, yet the one knowledge is not the other, since a different temporal mode constitutes a difference in the content of the judgment.

Similarly in relation to his feeling or will. In regard to one and the same event God's will is at one stage an intention to effect this event in the future or to allow it to be effected. At another stage it is a willing that wills it for the present moment itself and thus becomes the cause of its occurrence. And there follows then an activity of feeling which takes pleasure in the event as something that has happened. Thus again: God himself is subject to a multiple change; but this is not one that brings him into conflict with himself, making him more, or less, perfect, since it is precisely that change which preserves him in constantly the same perfection. Nor

87

is it a change which would suddenly and immediately bring about a difference, since it continues forward infinitesimally and with an always completely uniform velocity. For what God wanted two days ago for two days hence is what he wanted yesterday for one day hence, and it is what he wants as present today and will take pleasure in tomorrow as something that happened the day before, the day after tomorrow as something that happened two days ago, and so on, from eternity to eternity.[80]

3. It is obvious that this change in God must be of influence on the creatures existing at any given time. What God creates today will be conditioned, in relation to what he created yesterday, by a modified creative cause. Only this could give being to this time, and without it the event would not have been possible. It will be a condition which somehow modifies the created things themselves together with all of their properties. If one considers this properly, then one sees that already a kind of change is introduced into all created reality, and it accompanies all perseverance and change of every other sort and proceeds uniformly in infinitesimal differences.

4. This observation can be extended further still. The change in God was revealed to us through the empirical fact of change in general. There is, however, more that can be inferred from this fact, namely that there would be a change in God even if he had taken it upon himself to create a maximally unchanging world or even to create no world at all. For just as now different actualities follow on from one another, so in such circumstances would different possibilities. For it would after all still be possible for God to want the same events for earlier or later or for him at some earlier or later time to want the same as present. Thus in place of the missing positive operations of thought there would exist even in such circumstances a continuous series of negative decrees, all of which would equally exclude something's taking place.[81] This series would be without beginning and without end, and it would have a length exceeding every finite measure; but it could not have this length without differences for the individual points within it and these differences would in their totality yield a continuous infinitesimal change. Such a change would therefore indubitably obtain for the divine reality even if there were no other reality that were subject to change. But if such a reality exists, then this change necessarily occurs in it, too, and if this reality would perdure in a maximally uniform way, this change would constitute the precondition for a length and for a greater or shorter length of duration. Wherever there is a greater and smaller, there must also be a plurality of

parts, and the parts could not be a plurality without being differentiated from each other.

5. From all of this I believe that I have established that it must hold of the real as real that it is temporally extended. This lies in its concept. And by this I do not mean that it must hold of the real that if it is to be true with *one* temporal mode, then it must be true with a continuity of such mode. Rather, I affirm that that object which is true in such a continuity of temporal modes must, in so far as it is a thing, be subject to a certain infinitesimal change which proceeds uniformly and infinitesimally during the whole time of its existence, so that considered from this point of view its every later existing moment is ever more different from its every earlier existing moment in proportion to the length of its existence, just as in the case of a straight line in space every following point lies further from the starting point than any intermediate point which precedes it. This is what I used to refer to as differences of *transcendent* time. They have this peculiarity: that if two of them are specifically different then they cannot be such as to be affirmed with the same temporal modes.

6. One can ask whether the concept of what is real, as we in fact possess it, does not include something of this concept of transcendent time, and if so whether it comprehends only its most general marks or something more. One might believe that these most general marks are inseparably accompanied by certain further determinations. For the concept of transcendent time is the concept of a continuum of which it holds that every part allows itself to be differentiated into indefinitely many parts, each specifically different from the others, and of which it also holds that every boundary can be thought only as boundary of a continuum and thus not separated from the thought of indefinitely many parts of which each has its specific peculiarity.

Indeed it seems to me that the concept of transcendent time is contained as to its general marks in the concept of what is real. But more: it seems to me to be an irrefutable consequence of this that the latter is therefore also given as the concept of a continuum and therefore of something comprehending infinitely many differences of transcendent time. *Not, however, anything further*, and thus in particular not any absolute particular and specific determination of transcendent time, and, connected with this, the presentation of different and still more of infinitely many specifically different absolute temporal determinations. The case seems to me to be analogous to that where I have the general concept of a spatial point in abstraction from any absolute specific spatial determinations yet still bound up with the general concept of a spatial continuum whose boundary it is.[82]

7. Even in relation to what is psychically real our presentation of transcendent time is given only in this generality. It is as psychically real that we grasp ourselves, and the psychically real is not, like the physically real, given to us in a presentation that extends through different temporal modes. Wherever we have a so-called external perception, for example see or hear something, we perceive something as persisting or as changing. This happens in such a way that, while we present something newly registered by the senses as in the present, what had been earlier registered we present to ourselves with temporal modes of the past ranging in continual transition from the present to a more or less distant past.

Now we have at the same time as an external perception an inner perception that is included in the same act. 'Sight', as Aristotle said in this regard, 'relates, on the side, to itself'. And thus we perceive ourselves with evidence both as presenting something present and also as presenting something more or less recently past. Does the secondary object then, like the primary object, after it has appeared to us as present appear to us also as past? Analogy seems to speak for its doing so, and one might hold that if it did not, then inner perception would be unnoticeable. For would it not then be limited to a single point in time and as far as its object is concerned lack all extension, where in fact even a too small extension already stands in the way of noticeability? Indeed one could go further and say that, just as a point cannot exist for itself but only in connection with something whose boundary it is, so also it could not be perceived for itself but only together with something whose boundary it is.

8. This assumption that *inner* perception, like outer perception, would have a proteraesthesis too, of which perception of something as being present would only be the concluding boundary, is however subject to great difficulties. In such an inner proteraesthesis an earlier perceiving would have to appear to us as earlier, but as directed to something as if it were present. We would thus have to present also this recently existing former external object both indirectly as in the present and directly as having recently existed. Moreover, since also the inner perception which now appears to us as past was directed to an inner perception as recently existing, this inner perception too would now have to be given to us indirectly, and it would have to be given with the same temporal

mode with which it earlier appeared to us directly. And if this were so, then its primary object would have to appear to us doubly indirectly with the temporal mode of the present. Things would have to go on in this way. And because there would be given to us in the proteraesthesis a series of perceptions following upon each other continuously (not only *one*, as in the perception that is directed to the present psychic act with which it comes to an end), this would lead to an endless, indeed to an endlessly endless complication, in fact to a complication infinite to the infinite power, which can after all hardly be allowed to be accepted as a possibility.[83]

9. Or can this consequence be avoided? Perhaps someone will say that the inner perception which appears to us with a temporal mode of the past reveals to us the primary object also in the same temporal mode. The temporal modes of an earlier inner perception which the present inner perception shows to us would then be correspondingly modified: the present inner perception would show us that inner perception which stands to the earlier inner perception in the past in the same relation as that in which it itself relates to what is present, namely as lying in a more remote past and not in a past like that in which it itself appears. In consequence of this the infinite complication would not arise, since what appeared indirectly in earlier moments would often coincide with what appears directly in the present.

I hold this solution to be impracticable. The earlier inner perception did after all show something whose primary object appears in part with the temporal mode of the present and in other parts, following on therefrom, with temporal modes of the past distancing themselves infinitesimally from the mode of the present. If, now, it did not appear in the proteraesthesis as directed, in its individual parts, in the same ways to the same object, then it would appear as something other than what it was. In other words it would appear falsely, and since the inner perception of the present boundary is the perception of something that — as lies in the nature of the boundary — is co-determined by the nature of what it bounds, then not even this inner perception could be correct, either. But if there is anything that is unacceptable here, this is the assertion that the inner perception which accompanies our psychic acts could be false in so far as it relates to these acts as they presently exist.

10. Let us therefore try to find another way out of the difficulty. Since we cannot after all agree to such infinite complication, we

have the task of proving that the objections against the idea that inner perception can do without a proteraesthesis are invalid.

The first argument, to the effect that proteraesthesis is required by inner perception because of the analogy to outer perception, is clearly not conclusive. Inner perception distinguishes itself also in other respects, above all through its character of evidence, and evidence could apply to an inner proteraesthesis as little as to an outer perception.[84] What holds of the latter, namely that it is lacking in evidence as much for the boundary of the present as for those of the parts devolving to the proteraesthesis, would therefore not hold of inner perception. If we now also take into consideration the unavoidability of the endless complication which must arise through proteraesthesis for inner, as contrasted with outer, perception, then given such differences of circumstances the inclination to risk an inference by analogy will surely disappear.

11. A more significant argument might appear to lie in the claim that the object of inner perception would be deprived of extension if there were no inner proteraesthesis, when even a very limited extension will already make a phenomenon unnoticeable. Here too however there is a simple answer. It is only those objects which are of their nature extended, such as bodies, that can become unnoticeable as a result of smallness of size or reduction to a point. Inner perception however grasps the immaterial soul.[85] And even this it grasps in a way that is not wholly without extension. For although the soul is, as a substance, unitary, it is yet in its accidents manifold, indeed continuously manifold, and this in two respects: on the one hand because the activities of sensation relating to spatially extended objects can be resolved into accidental pluralities corresponding to the parts of the given objects; and on the other hand because these activities, in joining a proteraesthesis with that boundary-sensation which experiences the primary object as present, thereby also become continuously manifold, and they become in this whole continuous manifoldness an object of inner perception.

If the argument had any sort of force, then it would still not apply, even given the existence of an inner proteraesthesis, in such a way as to require that like inner perception in general so also the evidence of inner perception would reach not merely to what is present but also to what is past, in so far as this falls within its proteraesthesis. For if what is present were unnoticeable for itself, then it could also not of itself become the object of an evident accepting judgment. But the extension of evidence beyond the present is quite simply impossible.1*The argument would therefore prove too much, and thus it proves nothing.[86]

12. This also suggests the answer to the third argument. Just as a point, whether it be spatial or temporal, cannot be set apart for itself because it is a mere boundary, so also perceiving cannot possibly exist for itself in a single isolated point in time so as to be reduced to a temporally punctual perceiving. But still, no earlier perceiving is at the same time as that occurring now; yet earlier perceiving existed up to that occurring now, and this suffices to make the present perceiving no longer appear cut free and isolated. For after all, everything else that is temporal exists only in a point without existing isolated in this point — in virtue of its connection in infinitesimal transition with what is past or future.

Something can however be grasped as existing in a temporal point without our grasping any preceding or following stretch of time, in spite of the fact that the temporal point itself cannot be without that which precedes or follows. For after all universals, too, cannot be in reality without individual differentiation, but they can be grasped without it. The peculiar character of what is universal, its indeterminacy, may still thereby however betray itself, as demanding a more specific determination in reality. This applies also in the case of the point whose boundary-character allows us to recognise its belonging to something that it bounds. Yet this boundary-character requires no specific magnitude of extension; it does not even require the existence of any specific second point in time, however close this might be to the first. And thus the grasping will perfectly well be capable of being limited to what is in the present; indeed it must be so limited already for the sake of its evidence, as also because of the infinite complication which would otherwise ensue.

1* The argument of Descartes then returns: we could have been created by God with our inner perception just as it now is, in which case the proteraesthesis supposed to be contained within this perception would be false.

VI. If there no longer existed any things, or only a timeless God, then nothing would ever have been[87]

Dictated 4 February 1915 *[T 20]*

1.

> 'For one thing is even God denied,
> that he make undone accomplished deed.'

Aristotle cites these words of the poet Agathon in his *Nicomachean Ethics* (VI, 2) and lends them his approval. And there ought to be few who do not spontaneously regard what is here asserted to be impossible for the divinity, as something contradictory in itself. Yet there are many who still believe in a beginning of the world and of everything temporal. For it is commonly taught of God that he is nothing temporal, in that there is in his life no course of events and no before and after. It is thereby expressly denied that eternity would be a time distinguished from other times only in that it has neither beginning nor end. The world, which according to them is supposed to have a beginning but no end, could therefore, if it were pleasing to God, also cease to exist as it had begun, for it is only because God's omnipotence sustains it incessantly that it does not sink back into the void.

I want now to provide the demonstration that if these views — which in one very essential point I do not share — were correct, then it would be impossible to maintain the proposition that God could not undo what has occurred.

2. To show this I draw attention above all to the fact that everything that is past or future is this only in so far as it stands in a certain relation to what is present. All knowledge of what is past or future is therefore strictly speaking knowledge of something present as standing at a certain distance either forward or backward from the thing in question.

This will be quite particularly easy to see when one considers that what has been has in every case been at a precisely determined temporal distance from what is (for example at a distance of one hour or one year). An indeterminacy in this respect, as in every other, is quite simply excluded. And the same holds also of what is to be. It, too, will be at a determinate temporal distance from what is now. If, however, everything that is temporal were annihilated,

94

would it still be possible to speak of something 'in the present'? Certainly not. For then also there could not have been anything that was an hour ago, a year ago, or any determinate time ago. It is clear, therefore, that in case of an annihilating of everything temporal there would not only no longer be anything present, but nothing would have been in the past, either.[88]

3. In order to get used to this initially strange result, consider the following question: suppose that the opposite were true, i.e. that even after the annihilation of everything existing and proceeding in time there would still have been something in the past. Would it then be such as to have been always equally long ago in the past, or would it rather become always longer and longer ago in the past?[89] To accept the latter would be to accept that time is still continuing in its passage even though nothing temporal would any longer exist, which is quite openly absurd. The former would, however, contradict the nature of what has been, as of everything that is temporal, that it demands a continual uniform progression. What Schiller said about time, that its stride is threefold — the future drawing hesitatingly nearer, the now flying past quick as an arrow, the past remaining eternally still — is entirely false. Every instant of the future changes in its distance from the now just as quickly as this passes away; every instant of the past distances itself in just the same proportion, and becomes ever more remotely past as every instant of the future approaches nearer and stands apart from the present in ever smaller degree.[90]

4. If, now, someone asks me what it is that I object to in the traditional doctrine of God because it would deny the proposition that even God could not undo what has occurred, then I answer: I do not deny, as does the doctrine in question, that even the divine life has a course and shows a before and an after and a change.[91] If it did not, then God could also have effected no change. Already Aristotle, who was caught in the error of believing that no change at all is conceivable in the case of God as immediately necessary being, was conscious of the great difficulty of making intelligible a change that would come out of what were completely unchanging. He assumed as proximate effect a completely uniform change, in fact a circular motion. This, as the circular motion of a sphere bedecked in a specific way with stars, is supposed then to generate a greater multiplicity of change through its influence on the other spheres and on the lower elements. But even a moderately probing criticism is enough to show that all of these mediating links would not be sufficient to exclude chance, which was after all Aristotle's

intention. The highest celestial sphere is supposed to be bedecked with stars — this already takes away the desired uniformity from the motion of this sphere. If one asks why a given individual star in its continual motion is now precisely here, rather than there, there is no other answer than: 'Because earlier, for example one hour ago, it was at a certain other point.' But the question then repeats itself for every point and leads to a *regressus in infinitum* lacking any sort of clarifying force.

5. Those who adopt the view that there is no progression in the life of God say that for God everything is as if it were in the present. What they do not see, however, is that if from the non-temporality of God one draws the conclusion that for him nothing would be past and nothing future, then one must draw equally the conclusion that for him nothing would be present, either.

6. Some, who took into consideration the proposition that truth is a correspondence of thought with thing, detected a difficulty in the fact that there are true sentences which relate to what is past and to what is future, where one cannot after all say that there is here given something outside the thinking with which it could correspond. They had however not seen that whoever asserts something as past or future, always asserts something as in the present — something that is accepted in the one case as later than that to which the other time is assigned, in the other as earlier — and that this exists outside the thinking. Someone might find this insufficient and appeal to an analogy with juxtaposition in space, where, if something is above another thing, then this other thing must be below the first thing. Thus he might demand of succession in time that if one thing is later then the other must be earlier and conversely — but then he would demand the impossible. That which is earlier and that which is later cannot exist together.

7. The relation of earlier and later is in a certain sense to be compared with that between thinker and what is thought: the thinker exists in the proper sense, but what is thought exists only as being thought. If the latter, too, were a being in the proper sense, then even what is contradictory would be able to exist. Thus also if something is past or future then there is always something present and this is accepted in the proper sense as something earlier than what will be or as something later than what was, and what is past or future itself exists in the proper sense as little as what is thought. One should not however misunderstand what I have said here as though I were affirming of what is past and what is future that it

exists in the same improper sense as what is merely thought. There obtains, in all probability, a difference between something past and something thought as past.[92] Temporal relations are not psychical relations to an object, even if the two have in common that in both cases the terminus of the relation does not, like its fundament, have a being in the proper sense. We might call them historical relations, taking the word 'historical' so generally that it includes also what will happen in the future. And this, too, if our deliberations are correct, is to be said of such temporal relations as of psychical relations to an object, that it is in each case the fundament alone that is presented and accepted *in modo recto*, the terminus *in modo obliquo*. What is accepted *in modo recto* is always what is present.

8. Perhaps someone will object that we speak of earlier and later not only in relation to the present but equally in relation to what is past and future. Thus we say that one thing has been earlier or later than another thing, or will be earlier or later than another thing. What, then, is accepted *in modo recto* in cases such as this? My answer is that here the relation of what is present to what is non-present is thought twice over, once to something further away, once to something less far away in time. And both are accepted only *in modo obliquo* in the manner of what is historical, but not in such a way that they would be accepted with an exactly similar *modus obliquus*: as in general, so also here, there obtain the most manifold of differences.

9. Stumpf speaks in his *Tonpsychologie*[93] of a type of manifold that is grouped always around a certain unity as centre. One such manifold would be that of time-differences, where the point around which all differences are grouped is that of the now. And the case would be similar in regard to differences of space, all of which are grouped around the here. Yet I fear that Stumpf, in allowing space and time to appear in this way as analogues of each other, has like many others gone too far. For in the case of time, which exists only in a single boundary of the continuum, the point in question is exceptional in a material sense. In the case of space, however, there is no point which distinguishes itself materially in the same sort of way. It can only be said subjectively that a certain part of space has this advantage over others, that it alone is given to us phenomenally in our sensory intuitions. This subjective distinction has therefore a quite different significance, and a further difference lies also in the fact that we have to do here not with a point, but with an extended spatial stretch. Certainly there have been some who have still wanted to distinguish within this phenomenal region

a certain exceptional point and have referred to it as the position which our ego would enjoy in space. People do after all commonly say that they see something in front of them, that they hear something on the right or left, or from below or above. If, however, we were to grasp our ego as located, then we should grasp it thus in inner perception, and because inner perception is evident, it would have to be the case that we would notice in inner perception every displacement of our ego in space, but this is not the case. If one considers the phenomena as actually given, then one will see that what is at issue here is merely the fact that, given the limited extent of our spatial intuition, certain parts will naturally lie more in the centre and others more on the periphery of our visual field. In other senses, too, there are certain locations which have for us particular significance, so that others appear to us as less important. This holds in a certain sense of the impressions which we enjoy through the sensory organs of the head, and again as regards the impressions of sight, of the place of clearest and best localised sight.

It must however be added here that our phenomenal space has significance for us above all because it makes possible (at least relative) knowledge of real space.[94] This is the case to such a great extent that we regard the relations of phenomenal space not so much as such, but rather as indications or reference-markers for our judgments as to the material world. Thus we confuse and confound the associated thoughts relating to real space with moments of phenomenal space. Since, now, our body is located in a specific position of real space and everything that we otherwise find out about what exists in real space is made known to us through influences on the organs of this body, it is quite natural that the position of space where our body is to be found, and then also the neighbouring regions, should be of exceptional interest to us and that therefore not so much a point in real space but certainly a region of relatively limited extent is to be designated from the standpoint of our subjective interest as one about which everything else in space would be arranged.[95]

10. One sees here how much there is lacking a more than superficial analogy with time. And such an analogy is lacking in particular in that, in the case of spatial juxtaposition, where all parts equally exist in the proper sense, the differences thought by us are none other than object-differences, whether we have of them absolute knowledge or only knowledge of a relative sort. In the case of time, in contrast, where only one single point exists in the proper sense and all others exist in the improper sense (and even these do

not exist in the same sense, since the *modus obliquus* changes for every point according to direction and magnitude of its distance from the now), it is above all these differences in the *modi* of being in the improper sense which make themselves felt in our experience. Indeed when the matter is precisely investigated one finds that the temporal differences in our presenting, judging and in our emotional activity, are in the first place restricted entirely to these modal differences. Thus when we present a duration the object seems as such to be completely unchanged and to differ only in that —in addition to its acceptance in the proper sense—it is accepted with a continuity of acceptances in the improper sense and with a continually varying *modus obliquus.*

11. It is necessary to draw attention to one further similarity between historical and psychical relations. In regard to the *modi obliqui* of psychical relations it can happen that something is presented in this mode in a way that is either closer to or more distant from presentations *in modo recto*. This occurs if I present to myself a thinker who is himself presenting a thinker and indeed a thinker of a thinker — and the process can be several times repeated. Only one thinker is here presented by me *in modo recto*, all others in a *modus obliquus*, but they are not all presented in the same way, since the one *modus obliquus* stands closer to the *modus rectus* than the others. Something to some extent similar to this is to be found, now, in the series of presentations where different things (or indeed the same things) are presented and accepted with different historical *modi obliqui* progressively receding from the *modus rectus*. In this case however the series in question is continuous, where in the former case it was discrete, and this implies the further difference that here something can be thought in a remoter *modus obliquus* with elision of the intervening *modi obliqui* and in combination with the *modus rectus* alone, where in regard to psychical relations with a remote *modus obliquus* not merely must something be included in our thinking *in modo recto* but also something in each of the intervening *modi obliqui.*

12. I have said that, as the linguistic expressions themselves suggest, where we think something as enduring we think of it as persisting completely unchanged. Thus we often say that something has been preserved completely as it was. Common sense here gives testimony to the fact that when we think simple duration there are manifested no object-differences between that which was and that which is. And the testimony of the most important philosophers of antiquity, of the middle ages, and of the modern period agrees with

this. One can compare in this respect Aristotle among the Greeks, Aquinas and Suarez among the scholastics, and Leibniz among the moderns. The acknowledged difficulty of specifying what time is is connected with this.

However I believe myself to see clearly that everything that endures, however little it changes in ways that are available to our presentation, must yet still be subject to a transcendent continuous change and that without such a change it could not be subject to differences in length of duration. What someone like Suarez says about cases where one is supposed to be able to speak of an external but not of an internal length of duration, or indeed of external differences although the inner length remains the same, is demonstrably absurd.

If one asks in what this objective transcendent change consists, then a more profound investigation will lead to the result that it rests on the fact that what endures is subject to the unceasing influence of that first principle which is caught up in a constant uniform change. This principle must then stamp everything that it sustains in a given moment with the same specific character of the corresponding moment of its own existence. It follows from this that the temporal object-differences could be available to our intuition only if this first immediately necessary being were itself open to intuition. So long as this is not the case, our intuition of temporal differences is limited completely to differences in the thinker of what is temporal in so far as he presents and judges one thing *in modo recto* and other things in historical *modi obliqui*.

VII. Material and modal differences in what is temporal

Dictated 7 February 1915 *[T 22]*

1. That time belongs to what is real is certain from the outset to those who know that everything that is in the proper sense is real. It is however easy to see also that what is real must in some way be substantial, for then it will relate also to what is accidental. The converse does not hold: if the real were something accidental, then its substance would not have to be affected by reality. We know of substance that it can exist as universal just as little as can anything else, i.e. that it cannot exist without specific and individuating differences. It is conceivable at the outset that it has a multiplicity of interrelating series of differentiations, since this is the case also for the accidents.[96] And it is conceivable also that among these series are relative determinations, for it is by no means inconsistent to suppose that something is substantial and relative, if one conceives that which is relative so widely as to comprehend everything which is such that a presentation *in obliquo* belongs to its presentation *in recto*, as is the case for example for a thinker and for something that is continuous.

2. Yet however many and varied the differences might be that a substance has, nothing seems to be more certain than that we are not able to know these differences and to present them intuitively. Already Pascal noticed that we are able to distinguish our own selves from other thinking subjects in no other way than in relation to accidental determinations, as for example according to differences of our beliefs or of desires. In regard to what is spatial it would be conceivable that we somehow knew of its substantial differences only if differences of place were such. Then, however, changes of place would have to be conceived as substantial changes, which normally does not occur. Even then however what was said about lack of knowledge of substantial differences would remain secure for everything mental, and in particular for our own selves.

We have, therefore, no concept of (psychic)[97] substance other than that of substance in general. Hence this will be the only concept that we have of what is temporal, since this, too — as something continuous having one dimension and existing in a boundary — presents itself to us only in connection with what is known to us only according to its most general concept. We know, certainly, that it must be somehow specifically different from what it bounds, but we cannot specify the peculiarities of either the one or the other. We can say only, on the basis of what lies in the general determination *thing*, that there is given here a relation of before and after.

3. This before and after has the peculiarity that one cannot conceive it except through a multiplicity of presentations, and thinking a succession distinguishes itself in this regard from thinking a juxtaposition. This does not mean, however, that a plurality of modes of presentation were not also in a certain way required for thinking what is juxtaposed, since for everything that is relative there is presented not merely something *in recto* but also something *in obliquo*. But it is undeniable that for the presenting of what is temporal there is added to this multiplicity of ways of presenting also another.[98] For what is juxtaposed is a matter of juxtaposition of what *is*. If, however, something is, but is after something else or before something else, then this something else is not, but rather was or will be. The acceptance of something that is does not involve the acceptance in the same way of this other thing with which it stands in relation, but rather in another, indeed continuously other, way.

We are well aware of this continuous manifold of ways in which we must think the temporal as such, but we are not aware of the substantial differences[99] between what is thought with the one *modus* and what is thought with the other. And if we ourselves show in our thinking a manifold of modes of presentation that is of only limited extent, we do nevertheless see that this manifold could vary in such a way as to transcend every arbitrarily imposed limit. The variation in question is, after all, completely uniform.

This variation brings with it the differences of present, past and future, which are not differences of the object since every object can be thought not only as past but also as present and future. It points however to a change in the things, and more particularly in the substances, a change which is to be thought as continuously uniform and as taking place for all substances with the same velocity but which is not revealed to us intuitively in its specific differences.

4. This change is to be conceived in its generality in no other way than that the first and of itself necessary being, when it is a cause of change, cannot itself be changeless but must in its change be absolutely harmonious and uniform. In that this first being unceasingly conditions and sustains everything that it causes, it does this according to the continuously different moments in the

course of its own life. And so the dependence-relation, as dependence-relation of something that is in a certain sense continually other, is itself continuously changing, and it is this which makes up the temporal difference of that which exists at the same time in relation to that which has existed and will exist at a certain equal temporal distance therefrom.

I do not need to repeat that all these temporal differences of created being are, though themselves substantial, none the less relative to the first immediately necessary being, and they are as far as we are concerned completely transcendent. For them to be intuitive to someone he would have to behold the gradual course of the immediately necessary being. And as certain as it is that this is to be thought as God's essence, such a one would therefore participate in the intuition of God.[100]

VIII. The temporal as relative

Dictated 13 February 1915 *[T 24]*

1. It is above all a matter of setting forth the meaning of expressions like 'present', 'past', 'future'. These are clear for everyone, but in no way distinct. All are relative determinations in relation to the speaker. Present means: at the same time as the speaker, past means that the speaker is later, future that he is earlier.

2. To elucidate further the peculiarity of these *relativa* it is advisable to look first of all at what is common to everything that is relative. For each relative, there is presented (and possibly accepted or rejected) something *in recto* and something *in obliquo*.

3. From time immemorial one has distinguished a number of classes of what is relative: intentional relations, causal relations, and comparative relations.[101] Each of these classes manifests a plurality of species. The *intentional relation* is different in the case of presenting, accepting and rejecting, loving and hating and manifests still further differences subordinate to these.

To the class of *causal relations*[102] there belongs everything that somehow conditions the being of a thing or is conditioned by it. Thus this class includes also parts as such ('partial causes'), since the whole could not be without the part. One has often called this the material cause. The class of parts also includes the substance which as subject is contained in its accidents as part, and it also includes every accident that underlies another accident. It does not however, when precisely considered, include a logical part, since the universal is individuated in the thing.[103] Included further is that which is continuous ('continual cause'). In the case of what is spatial, continual causation is (where we are not dealing with contiguity) a mutual causation between a boundary and what it bounds. The class of causal relations includes also the efficient cause, to which is related that from out of which something else develops on coming into existence. Some would also like to include here that which has the power to receive something in itself, as a precondition of realisation, while others might want to count this among the partial causes.[104] A special mention is also required of what takes place according to the law of inertia and other kinds of persistence.

Those things which stand in *comparative relations* vary according to the *tertium comparationis* and in relation to agreement or difference, the latter culminating in opposition. If we count the continual relations as among the causal, then we shall have to reckon the relation of contiguity among the comparatives.[105]

It is evident in regard to the comparative relations that they are a matter of a relation between two *realia*, and they presuppose a reference to our activity of comparison. If there was no one who could compare, then there would fall away also comparability and the possibility of being found to be similar or different. Should one therefore perhaps say that the *comparativa* stand closer to the intentional relations than do the relations of causality? Indeed the fact that the relation of being earlier as also of being later exists in the proper sense only in its fundament will remind us still more of the intentional relation.

Certainly someone could say that something similar was the case even for the causal relations where cause and effect border on each other temporally. Taken precisely however this is not correct: much rather must efficient cause and effect be at the same time,[106] at least in the sense that they coincide temporally as ending and beginning. If the production of effects goes on for a certain period of time, then the full simultaneity of effect and cause is still more apparent, and when a completely rigid body is displaced, the motion of the more remote end of the body takes place fully simultaneously with that of the parts lying nearer to us but mediated through a continuity of the intervening parts. If fundament and terminus are both real for other comparative relations then this is something that occurs also for certain activities of thought, as for example when something is accepted with evidence. For then the terminus of the relation must after all be in the proper sense, no less than the fundament. To bring the causal relation in closer approximation to the intentional relation one might point rather to the fact that, just as thinking posits something only in the thinker and not in that which is thought, so the causal relation posits something only in that which is effected, not however in that which has the effect. The comparative relation however posits in both nothing specifically new. Only of he who makes the comparison himself can it be said that the comparison constitutes in him something new. The act of comparing belongs however to the activities of thought, and this leads back again to the idea of a special affinity of the two classes of causal and intentional relations.[107]

4. On the basis of the simple relatives there are built up more

complex ones. For example when someone thinks a thinker of something, or when someone says that Caius is bigger than you imagine him to be. Here the intentionally relative combines with the comparatively relative.

5. The mentioned classes of relatives reveal a significant difference. Certain cases are such that if the fundament exists then so also does the terminus and these could be designated as *relativa* in the narrower sense. Thus 'Caius is taller than Titus', which means the same as 'Caius is taller than Titus *is*'. Not however in other cases, as for example 'Caius is thinking of a centaur'. (Instead of *relativa* one could call these also '*quasi-relativa*'. The similarity with relativa in the narrower sense consists in the fact that in both cases besides something *in recto* something is presented also *in obliquo*.)[108]

6. Let us turn now to the case of those *relativa* whose elucidation is our specific task. It is at any rate without doubt that we have to do here with comparative relations. In regard to what is present we have to do with a case of an agreement in which the speaker stands to that of which he speaks, and the *tertium comparationis* is the temporal determination. That which is called 'in the present' exists with the speaker at the same time. As the latter now *is*, so also is the former. It is clear, too, that one has to do here with a relative in the narrower sense. Not merely the speaker but also that which he refers to as present must exist, if the former is to be at the same time as the latter.

7. The case is in a peculiar way different in regard to what is past and what is future. Here, too, we have to do with comparatives, but in the case of what is past the speaker is later than that to which he relates as past, and this past *is not*. One cannot even say that it is earlier than the speaker, but only that it has been earlier than the speaker, and of the latter that he is later than it has been. Things are similar in the case of the relation to what is future. Here the speaker is earlier, however not earlier than something that is, but earlier than something that will be.

8. What is now to be said of this difference between *is*, *was* and *will be*? Does it materially separate out what is from the rest? Does a real attribute, different from those which the speaker has, come thereby to be ascribed to that of which he speaks?[109] And is this, if not absolute, then at least a relative attribute? The first will hardly be allowed to be maintained; rather perhaps the second, since as already said, when dealing with present, past and future we have to do with a relation, indeed with a comparative relation to the

speaker. However we have said already that this relation, so far as past and future are concerned, is not a relation in the narrower sense. What is past and what is future are therefore not accepted as is the one who is set in relation to them. If this latter is, then the former are not, and therefore also not with any special material attribute which the latter lacks.[110] Moreover, the expression 'is' contains for the speaker himself no real predicate, and if in the 'was' and 'will be' there is brought to expression something different from the 'is', then still it is something coordinate therewith, which indicates, too, that even here there is expressed no material attribute.

What however is it with which 'is' as expression is coordinated? Here it would be above all 'is not' which one might mention. Yet simple negation certainly does not produce the same as 'was' and 'will be' — even if what was, in so far as it was, no longer is, and what will be, in so far as it will be, is not yet. So, if it is said of something that it was or that it will be, should we count this among the affirmations or among the negations?

As was established already, the determinations of present, past and future are relative, more particularly comparative, determinations of the speaker in relation to that of which he speaks. In the case of the present this is a matter of agreement, of sameness in regard to time. In the case of past and future it is a matter of differences in regard to time, in both cases of differences of something that is in relation to something that is not. It is, in the two cases, a matter of differences in opposite senses, in that what is is characterised in the one as later, in the other as earlier. If one now accepts the speaker, who is, as later than something or as earlier than something, then here, as everywhere where we have to do with a relative determination, there will be besides what is presented and accepted *in recto* also something presented and accepted *in obliquo*. This is the case even where we are concerned not with relatives in the narrow sense but merely with what is quasi-relative. Hence we can now say: if something is presented and accepted as past or future, then it is presented and accepted *in obliquo* as a result of the acceptance *in recto* of something which stands to it in the comparative relation of the earlier or of the later.

9. It remains for us still to investigate what this earlier and later might mean. This will best be served by means of examples of cases where they are given to us intuitively. Most of us will be inclined to single out spatial motion as such an example: the case where, as one says, we 'see' that something is moving in front of our eyes. Yet

precise consideration reveals that one does not properly *see* a motion here at all. The thing that moves in front of our eyes successively stimulates different loci of the retina and each stimulus of a different locus yields a locally different sensation which, for as long as it exists at all, is unchanged in its location, i.e. yields more a sensation of rest than of motion. It is merely that these for us unnoticeably small periods of rest present themselves to us in their multiplicity as a case of succession. And similarly in the case of the sounds that we hear in a melody or in a speech. The musical tones, the words and syllables in the speech, appear to us as a succession, as a manifold of earlier and later. We ourselves as ones who see and hear apprehend ourselves as something that is; the tones of the melody, the words and syllables and letters of the speech, are presented and perhaps also accepted, the one as present, the others as past, yet this latter would be without logical justification were it not that we who perceive them are related to them *in modo obliquo*. The one is then however accepted as something that is experienced as present, the other as something that is experienced as pre-present and as pre-pre-present, i.e. it is accepted as more remotely past or as earlier than the present and still earlier than the present, and so on.

10. It is well known that we are not in a position to differentiate distinct parts in a continuum beyond a certain smallness, and still less does our capacity extend to the differentiation of a single boundary within a continuum. This does not, however, stand in the way of our noticing and clearly apprehending the character of the continuum in general. And we can thereupon establish deductively that it is made up to infinity of smaller and smaller parts, and again, that these touch each other in null-dimensional boundaries, points, which would not be conceivable if there were nothing which they would bound. In the case of a one-dimensional continuum these boundaries can be internal boundaries in two opposing directions and are then points of connection. But they can also be internal boundaries in merely *one* direction in relation to that which they bound, and in the other direction be external boundaries. They are then separating points, actually not one, but two of half plerosis which coincide. And again it can happen that a boundary bounds on only one side without it being the case that another one-sidedly bounding point is present which coincides with it.

11. Our inner perception, which perceives with evidence, is entirely restricted to a single now and presents us with certainty only with one boundary which is in the present. Yet still, it

manifests this also *as* a boundary, i.e. as being by virtue of its own nature the boundary of a continuum that it bounds. And this continuum, because we are dealing with what is temporal, can be conceived only as earlier or later or also as both earlier and later. Thus in spite of the limitation of inner perception to the one point it is not only that it presents and apprehends this point *in recto*; it presents and apprehends also *in obliquo* a temporal stretch of which the point is a boundary. If one asks as to the length of this stretch that is presented and accepted *in obliquo* then one has to answer: it has no determinate length at all. And this is in conformity with the fact that we do not appear to ourselves in inner perception as individually determined and as specified in all respects. No one can say what the substantial, individual peculiarity might be, that differentiates him from other minds. This means however, since every individuation is given through specification, that we also do not know what it is that differentiates the species of our substance from that of other minds or spirits. We perceive merely that we are substances which have as accidents certain activities of thought and that we belong to the temporal continuum of a certain substance. There is however nothing here that could not be said equally of every substance, indeed demonstrably even of the Godhead.[111] We experience not even the very least about the substantial differences of a bounded thing which is presented and apprehended *in obliquo* as earlier or later.[112] We present it to ourselves as far as its substance is concerned, like that which is presented *in recto*, only as thing in general.

Although inner perception manifests to us this bounded thing *in obliquo* (with the exception of its boundary, which it manifests *in recto*), and manifests it to us without any specific length, still, it tells us that it must have a length and that, as for every length, ever smaller parts are given to infinity therein. Of these, some are nearer to the boundary, some are more remote, so that the boundary exists as later than all past parts, though at a distance that is less in the case of some of these parts, more in the case of others, and analogously also in relation to the future parts to which it stands as earlier to a greater or lesser degree.

12. What this more and less signifies can be best explained by means of a comparison with what is presented *in obliquo* in the case of the intentional relation. If a thinker presents a thinker who himself presents a thinker and so on in tenfold iteration, then the one who thinks a thinker *in recto* perceives the thinker thought by him *in obliquo*, but in such a way that the first is *in obliquo* in the

first degree, the second *in obliquo* in the second degree, and the last *in obliquo* in the tenth degree. One can therefore speak in regard to the latter of a greater mediacy which in a certain sense distances it further from being presented *in recto*. The case is quite similar in regard to that which was earlier and will be later. In the one case something thought by a thinker is as thought *in obliquo* at a greater distance from what is thought *in recto*. In our present case something thought as earlier than something else is set at a greater distance from what is thought *in recto* as later, or something thought as later than something else is set at a greater distance from what is thought *in recto* as earlier. Of course there remains this difference between the two cases: in the case of the thinker who thinks a thinker etc. the greater distance is measured by a number of discrete units; in the case of what is later that which is later than something earlier (which is in turn later than something still earlier etc.), or in the case of what is earlier than something still later (which will in turn be earlier than something later etc.), the distance from what is presented *in sensu recto* is increasing continuously.[113]

13. If we should thereby have succeeded in demonstrating the origin in intuition of the concepts 'later than something that was' and 'earlier than something that will be', then it will not be difficult to understand how something in the present can be thought of as at an arbitrarily large distance from something more and more remotely in the past or future. This will be able to happen in a way quite similar to the way in which we can, on the basis of our narrowly restricted spatial intuition, arrive in thought at extrapolations of spatial extension in all the three dimensions. For our spatial intuition, too, is limited, and as concerns the third dimension, that of depth, perhaps even limited to an infinitesimally small magnitude that does no more than to secure quite generally an extension also in this direction, without however there being phenomenally given even a single boundary lying either to the front or to the rear.[114]

14. There remains however one extremely significant difference between the spatial and temporal cases. In our intuition of what is spatial every point, in standing apart from another, shows itself with materially different specifications.[115] Just as something coloured shows itself as standing apart qualitatively from something differently coloured, for example something red from something blue, so also something spatial here shows itself as standing apart from something spatial there. Quite different is the case of something that differs as earlier and later. Distances which

concern material differences, too, may show up also in this case, as for example when there is noticed a transition from one tone to another in a melody, perhaps to a tone standing far apart within the scale. Yet such a material difference contributes absolutely nothing to the *temporal* distance and can indeed be entirely absent.

The utterances of both laymen and of distinguished philosophers provide testimony for this. The way the laymen conceives things is revealed in a number of our more common linguistic expressions. Thus one speaks, when referring to the opposite of a process of transformation, of perseverance, of a persisting, of something's being preserved completely, and one sees nothing contradictory in the assertion that something that is now was already at some earlier time and was just the same and in just the same way that it is now, and also that it will continue to exist as just the same and in just the same way that it is now, if nothing interferes. One speaks of a *continued existence* and that is very characteristic of the fact that one does not believe that temporal displacement is something analogous to spatial displacement. Otherwise one would have to speak not of a *continued existence* but rather of something's *going on existing*. It is commonly supposed also that something that has ceased to exist can be reconstituted in a later time in such a way that it would be exactly the same as it was and in every respect just as it was, even if not by human power still in any case by the power of God.

As concerns the view of the philosophers, the older Ionians conceived things exactly as our laymen. Did not Thales, Anaximander, Anaximenes, Heraclitus, Parmenides and Empedocles believe in the periodic return of exactly the old state of the world in such a way that every real difference would have been cancelled out? And does not this imply that distance in time was not counted as a real difference? Even Greek philosophy in its most mature state, as it is exemplified above all and at its best by Aristotle, did not think differently in this respect. In addition to motion he acknowledges also a state of rest and of remaining absolutely the same. That there appeared to him to be no real change through duration itself, even in the case of an arbitrarily long duration, is seen in the fact that he assumes only three sorts of transformation: of location, of quality and of substance, and in the fact that he declared that of these local change is the primary and is always such as to play a part in the other two. Sometimes indeed he seems to bring time into the closest of relations to motion in space, namely to that of the outermost celestial sphere. But he who pays precise attention will discover that he expressly protests against the view that time itself would be a motion, since it is much rather only something which is to be found attached thereto, namely the number of motion in regard to its before and after. He declares also that if the soul did not exist, then there would still be the motion with its before and after, but there would be no time, since there would be nothing that would do the counting. Since this addition proceeds always in the same direction, it follows that time is for Aristotle not so much comparable to the rotations of the heavens that are counted in the measurement of time, as to a rectilinear progression. All other utterances of Aristotle are in harmony with the thesis that he saw no sort of material alteration in persistence taken of itself alone. Thus when he discusses the ways of speaking of the common people according to which time could destroy something, bring something forth, he for his part denies it all effective influence. He speaks here also of the recurrence of exactly similar conditions, and surely no one will expect to see a contradiction in this. Thus he must hold a mere being later to be not a material difference, for he would otherwise have to believe in a return of time itself in the case where what is materially exactly the same is given for a second time. In the subsequent period, the period of ancient philosophy in decline, the Stoics reveal sufficiently well in their revival of the physics of Heraclitus, that they do not think differently in regard to time.

Nor do the investigations regarding time made by Augustine in his *Confessions* in any way betray any deviation. When he animatedly declares himself against Aristotle, this is still only in the erroneous belief that Aristotle had identified time with one of the movements of the heavens.

In regard to the middle ages it is sufficient to look at the *Disputationes metaphysicae* of Suarez, who at the end of scholasticism looks back on all his predecessors. Characteristic of his own conception is his thesis that a state of complete rest reveals no differences in length of duration when considered in itself. Only in regard to the motion and change that takes place during the fully unchanged existence of a thing does it make sense to say that a rest has endured 1000 years in the one case, one day only in another.

There are also utterances of the modern philosophers which could be cited as agreeing with those of older times. Descartes identifies the time of a thing with its duration, which apparently, according to him, allows it to reappear as something entirely the same.[116] Hume expects under the same material conditions the recurrence of the same, however large the temporal interval after which it should reappear. Temporal position appears to him therefore to make no material difference at all. Leibniz, like Aristotle before him, sees that the temporal differences do not themselves make themselves noticeable. Only what happens in the present, not the present as such, manifests differences of any sort. If the same thing should happen unceasingly, then there would be generated no difference at all. (Aristotle had already drawn attention in this connection to the interruption of wakefulness by a deep sleep: when consciousness returns, there is no noticeable distance between this first temporal moment and the last before falling asleep.)

15. What, then, do we have to say about the appearance of a temporal extension as opposed to that of an extension in space? The former belongs to a continuum that exists only according to one of its boundaries and is earlier than one part and later than another part of the continuum. Of these parts the one is not but will be and indeed will be removed to a different degree in every different boundary from the boundary that now exists. The other part is not, also, but it has been, and has been in each different boundary further removed from the boundary according to which it exists. The boundary, however, according to which it exists, does not properly remain, since otherwise it would after all reveal itself as extended. If an existing thing persists, then this happens through a continuously reiterated renewal. For just as the existing boundary is the end of the past part of the continuum and the beginning of the part that is to come, so also every earlier boundary has been the end of a past part and the beginning of a part that is to come, and every future boundary will be the end of a future part of the continuum and the beginning of a still more future part of the continuum. Thus in the case of perfect persistence, a thing appears to be in temporal connection with itself, its persistence however appears as a continually reiterated ending and beginning of one and the same thing in one and the same boundary. In the case of unchanging perseverance the same thing appears as later and earlier in so far as it itself appears to have been and appears to be going to be, and appears as standing to different degrees apart from itself as something past and as something future. It appears now *in modo recto* and continuously many times *in modo obliquo* and indeed in *modi obliqui* that are themselves continuously different.

16. This should not be taken to imply that, between the thing that appears to us as present and the thing that appears to us as

past or future, there would (or could) be found no material differences if the thing were to appear to us in its full individual peculiarity. This is, to repeat, never the case. As far as substance is concerned, we even grasp ourselves only completely generally,[117] and this has as consequence — since the accidents are co-differentiated by the differences of the substances — that we also grasp ourselves in our accidents only in general. Thus it could be that there are in fact material differences between what is temporally earlier and later as such, but that they do not appear to us in virtue of our incomplete apprehension of the substantial peculiarities of the things. The distances that do appear to us between earlier and later are not material. One understands also on the basis of the preceding discussions how it is possible that there should be a grasping of distances and of a continuous extension without grasping of material differences, where such a possibility is completely excluded in regard to space. In regard to what is spatial, where the whole continuum *is* in every one of its boundaries, every sort of distance would completely fall away if the material differences and distances should be removed. In regard to what is temporal, however, we have to do with a continuum that exists only in relation to a boundary: in relation to every other boundary it is at a distance from existence and is so in part on different sides, in part at least to different degrees, and thus also this distance-relation can be brought to presentation by presenting the fundament of the relation *in recto*, the terminus *in obliquo*, so that a continuum is thereby given intuitively in the absence of all material differences.[118]

17. Briefly: a temporal continuum is a continuum of such a kind that it exists according to a single boundary which belongs to the continuum even though the two remaining parts thereof do not exist and supplies for these two parts the boundary in opposite directions. It can be a continuum of this sort only because the boundary is so to speak a running boundary, a boundary in which one part of the continuum ends (ceases to be) and another comes into being, just as all other boundaries from which this stands apart in the one direction have themselves been running boundaries in which one part of the continuum has ended and another has begun, and likewise all boundaries which stand apart from this in the other direction will be running boundaries which end one part of the continuum and begin another. The boundary according to which the temporal continuum exists can be punctual, but it can also be extended, indeed in several dimensions. In the case of bodies, which do after all exist also as temporal, the boundary is three

114

-dimensional and a boundary of a four-dimensional continuum of which it bounds one part as final boundary, another part as initial boundary, while other boundaries of the same four-dimensional continuum which are equally to be thought of as three-dimensional have likewise bounded one part of the continuum as final boundary and have constituted the initial boundary in relation to another. Yet other, also three-dimensional boundaries will bound one part of the four-dimensional continuum as final boundary, another part as initial boundary. This four-dimensional temporal continuum can manifest itself as materially different in every one of its three-dimensional boundaries (each of which is such as to exist at the same time in all its parts), for example when a body moves from one place to another. But it can also manifest itself as materially completely undifferentiated in its various three-dimensional boundaries which are at the same time or were at the same time or will be at the same time in all their parts. Things are then such that, in so far as matters are revealed to our presentations, the same thing which has come to an end in any given past temporal point also begins again anew, so that it has renewed itself, something we express by saying that it has continued to exist, it has maintained itself in being, it has remained the same. In the boundary according to which it exists, so far as this is grasped in our presentation, the same thing ends and begins at the same time, so that it has renewed itself, something that we also designate by means of 'continues to exist', 'maintains itself in being', 'remains the same'. And in all boundaries from which this existing moment is set apart in the other direction, the same three-dimensional thing, so far as it enters our presentations, will end one part of the continuum and begin another. The distances which come to appearance therefore manifest themselves in relation to the measure of their boundary as being determined not by differences of material attributes but rather by the magnitude of the continuous amount of endings and beginnings. This is the case no matter whether a material variation is also given, or whether it is one and the same thing (so far as appearance is concerned) that has executed, executes and will execute this process of self-renewal in simultaneously ending and beginning with the given measure of frequency of iteration.

IX. On understanding the Aristotelian doctrine of time

Dictated March/April 1915

This item is based on five pieces dictated by Brentano on 11/12 March 1915 *[T 31]*, 31 March 1915 *[T 33]*, 2 April 1915 *[T 34]* and 4 April 1915 *[T 35]*, as also on one dictation *[T 47]* that has come down to us undated. Kastil did not reproduce these dictations in their original form, but produced a completely new text by bringing together the most important parts of each. [Editors' note]

A. The doctrine

Interpretations of Aristotle's doctrine on time have normally taken account only of what is said about time in the fourth book of the *Physics*. To become more thoroughly acquainted with his thoughts, however, the statements in his other writings must be consulted also. Thus already

I. the *Categories* and *Topics*

provide, where the categories are listed, examples of the when.

II. The *Physics*

The fourth book of the *Physics* deals with time in its second half. Aristotle raises here the question of the existence and of the nature of time.

1. *Does time belong to what exists or rather to what is not?* Much seems to speak for the latter. Past and future are not, only the present is. Time is supposed to be divisible, but its parts, namely past and future, are not, and that which is, namely the now, is not properly to be called a part of time, for otherwise it would have some measure.

What Aristotle conceives to be the solution to this aporia is not made sufficiently clear. Perhaps he believes that it is solved by stating that the present belongs to time as a boundary and that time participates in the character of motion which is, according to him, an imperfect actuality.

116

2. In regard to the *question of the nature of time*, he does not wish to give the answer of those earlier philosophers whose views he reports, that it is a motion, namely the rotation of the sphere of the fixed stars. This already for the reason that motions have degrees of velocity. Also motions are of different genera, time however of only one. Clearly Aristotle is here thinking of what one understands by time when reference is made to a certain hour, a certain day, a certain year, a certain century as 'a time', while the whole series thereof, the whole which they together constitute, is called 'time'. It appears to him as undeniable that all these determinations relate to astronomical motions and more particularly to the apparently most uniform rotation of the sphere of fixed stars. Still however time is not itself motion, even if it is not without motion, or rather without something that moves (changes). It is something that is able to be distinguished in motion and rest, whereby the word rest is to be restricted to the factually unchanging perseverance of what is capable of change.

3. *Time is the number of motion* (and rest) in relation to the earlier and later. This definition is not easy to interpret. One might make it more intelligible in the following way. In the case of a uniformly rotating disc, the velocity increases with the magnitude of the radius, but the duration of the motion is the same for both the nearer and the more remotely situated circumference. Thus we could allow ourselves to say that the continuous multiplicity of moments of the motion are the same in all these motions of different velocity. This then also holds for the multiplicity of moments of a rest or of an irregularly varying motion that would take place at the same time as this uniform motion (for example of a surface which surrounds a disc). These moments can be measured in regard to their earlier and later or in regard to the positions they successively occupy in terms of the quantum of the relevant parts of the continuous multiplicity of the uniform motion.

By number here, therefore, one is to understand what is counted or countable, and that is the successions which make up the whole duration of a motion or rest. One must thereby probably conceive the duration as divided into equal parts. In the case of motion we have to do with actual successions, with a sequence of actual changes, where in the case of rest we have to do only with a series of mere possibilities.

That Aristotle refers to time not merely as the measure but rather as the number of the motion becomes intelligible in the light of the usual numerical system of dating (on the 1st day of the 3rd month of the 2nd year of the 30th Olympiad) and probably also in reflection of a passage in Plato's *Timaeus*.

Time is to be considered as applied cardinal number, and every answer to the question 'when?' as ordinal number.

There almost certainly hangs together with this definition of time as enumerated multiplicity of succession of a motion (or of the possibility of such in a rest), the statement that there would be no time if there were no soul. Only the intellect can count. (According to Thomas the possibility of an intellect would suffice.)

4. *Time in the narrower and in the broader sense.* One can in a way distinguish time in every motion or rest, in so far as each has its own duration. In the narrower sense time is the number of the motion of the uppermost celestial sphere in relation to earlier and later. This motion is without beginning and without end. One says of motions and states of rest which have a beginning and an end that they are *in* time, in so far as they begin and end at the same time as a certain part of the motion of the uppermost celestial sphere.

Times in the wider sense and time in the narrower sense are continuous and therefore allow the differentiation into an infinity of ever smaller parts. The boundaries where these parts meet are however, as already said, not themselves to be called parts of a time.

The number as which time is to be defined is therefore a continuous number.

5. *Time and eternity.* Time is to be distinguished from eternity, by which Aristotle understands the existence, without beginning and without end, of such things as are *absolutely unchanging*. Here there is a succession neither of actually varying moments nor of possibilities. Such things are, he says, in their unchanging existence *not* measured by time. Timelessness of this sort is enjoyed by God, the immaterial movers of the spheres, and the heavenly bodies, these last however only in relation to their substance, not in relation to their place. Man's spirit, too, is unchanging in its substance and is for this reason described by Aristotle as eternal, though it has a beginning. This is connected with the fact that it was not, in the first place, the mind that came into being, but rather the human being, who is according to his form partially spiritual and, after the dissolution of his body, such that his mind remains in existence. (However, the fourth book of the *Physics* does not yet contain anything of this special doctrine.)

Truths such as those of mathematics are also reckoned among that which is eternal.

Time is therefore to be found, according to Aristotle, only where some thing is composed of actuality and possibility. Such a thing, for as long as it is at rest, certainly shows itself to be just as much unchanged as does the eternal (i.e. as that which is preserved not merely as a matter of fact but without even the possibility of change). Yet the two cases are still essentially different. In the case of what is at rest, we have a continuing to exist of something that could also come to an end, and thus the continually recurring realisation of a possibility. Its being at rest presents itself therefore as a continuous renewal of one and the same thing and thus it is also to be referred to as a flux, though one in which the inflow and the outflow are completely equal. The continued existence of what is imperishable is however not a matter of such inflow and outflow, for it is pure actuality, free of all possibility, and where there is no potentiality, so also none can be repeatedly actualised.

How are we to understand the view that the changeable, even if it is just now at rest, manifests a number of successions. This is shown by certain passages, for example in the books *On the Soul*, in which Aristotle speaks of a certain sort of passion that is not a corruption but an actualisation of something given in mere possibility and thus a perfection. In particular in regard to affects and noetic activities it is said in the tenth book of the *Nicomachean Ethics* and elsewhere that they are a passion of such a sort that they proceed without successive change and continue to exist in that one and the same thing is renewed through the permanent influence of an efficient cause.

The doctrine of Aristotle according to which not only does the first immediately necessary principle exist eternally without all succession, but also the created movers of the spheres and the heavens are to be denied every succession and temporality, has influenced metaphysics right up until our own century. (Compare, in particular, the Neoplatonists, the Fathers of the Church, the scholastics with their distinction of *aeternitas*, *aevum* and *tempus*. Also Leibniz.)

6. Because the heavenly bodies are unchangeable as to their substance but not as to their place, the question arises why Aristotle conceives their *motion as without beginning*. This is connected with the difficulty which he recognised of making intelligible how change could proceed from that absolutely unchanging first principle which is the Godhead. Since this

principle does not change, Aristotle's determinism requires that always the same should occur as its result. This is the case, he holds, if its immediate effect, so far as this is a motion, should be so regular that like would always take the place of like, and this would be possible only for a circular motion which also did not change in its velocity. Thus the motion of the uppermost celestial sphere, whose number is' time, must be of absolute evenness and without beginning and end.

Something similar follows for him also for the motion of the other celestial spheres and for their numbers, which Aristotle likewise considers as effects of absolutely unchanging spiritual substances. Through the mediating influence of the heavenly bodies adhering to these spheres upon the corruptible elements of the nether world, there then follow changes in the latter which differ in manifold ways.

7. *One-dimensionality.* The times proceed one-dimensionally and are comparable to a straight line in so far as the distance of the end-points always corresponds to the length of the time lying between them. This is so even if the motion in relation to which time is distinguished should manifest a curvature. Times, in contrast to spatial extension, exist only in a boundary, the now.

8. In relation to the *now* Aristotle raises a number of questions. Does it always remain the same? If one now passes away, does it pass away into itself or into another? Does one now touch the other?

He seems however to understand the *now in a two-fold sense.* Sometimes in the sense of the present instant, sometimes in the sense of a certain moment, thought of as the same whether it still belongs to the future or is already past and as being such that its distance from others does not change.

The now in the *first sense* forms the beginning and end of a time: end of the past, beginning of that which is to come.

The now in the *second sense* moves through time, it constitutes the unit in terms of which the number which time is, is measured. Time appears therefore as a continuous number of nows, whereby the expression 'number' is employed in a peculiarly extended sense in that the length of duration is called its number. It is the continuously multiple now. This second use of 'now' is not common. The expression 'sometime' would be more appropriate. One does indeed say of something that was, that it was in the present, of something that will be, that it will be in the present, but to speak of a past or future now would be disconcerting.

The *relation between the now and time* is identified by Aristotle in very doubtful fashion with that between what is movable and movement itself. As what is movable is, during its movement, materially the same but changes as to its concept, so also the now is, for the duration of the movement, materially the same though different as to its concept. He seems here to be aiming at the fact that of a continuous series of moments every one is successively in the present. But it would miss the point to interpret the present as the substance and the plurality of moments as the changing accidental determinations which this substance would enjoy. Not only substances but also accidents are in the present, and if a substance were to preserve itself unchanged while moving, then it would be simultaneously true of it in the midst of this movement that it was, is, and will be.1*

In relation to the question whether the now is always the same in every instant of duration, Aristotle also makes use of the comparison with a point by means of whose movement the geometer would construct a spatial line. The point would be essentially the same, but would change its spatial determination. Similarly the now, too, would be materially one, but would however change in that from something earlier it becomes something later. Here it seems that even the process that is adduced for the comparison is not sufficiently understood.

9. These are the essentials of the doctrine as it is to be found in the fourth book of the *Physics*. Aquinas in his commentary orders the questions dealt with by Aristotle in the following table:

1. Whether time is, and whether in it there is one now or many.
2. Time is not motion but not without motion.
3. Further explanations as to what it is in motion that time is.
4. In what sense the now in time is the same and in what not, and how it connects times and separates them.
5. In what way is time a number of motion.
6. What it means 'to be in a time' and to what it is that being in

---

1* Thomas Aquinas, who does not criticise what Aristotle says about the now, employs its identification with the unity of the moving substance as a means of clarifying the idea of eternity, which is supposed to be given where a substance is subject to no change. In place of the number of the motion there enters the unity. This, too, completely misses the point, particularly since a continuous infinitesimal change can be demonstrated for God himself, and since this change must then be counted as substantial if one excludes in God a composition out of substance and accidents.

time applies, and to what not.

7. Explanation of the concepts: now, then, already, once, suddenly.

8. That time is more cause of coming to be than of passing away, which takes place only *per accidens*.

9. Discussion of the question as to the existence and as to the unity of time.

III. *Metaphysics*

One passage in the 10th Book that is considered on a number of occasions by Thomas is 1053 a 8. The 12th Book deals with the impossibility of a beginning of time since in regard to what is prior to time one could not speak, either, of a non-being (6th chapter, 1071 b). Heaven and its motion must for this reason be eternal. (Motion can neither come to be nor pass away, and nor can time, for where no time exists there can be no earlier and later.) Such an assertion is not to be found in the *Physics*. Indeed the remark there, that if there were no soul, there would be no time, seems rather to involve the possibility of a beginning of time. Which of the two discussions seems the more mature?

IV. *De anima*

1. Very important remarks are to be found in the books *On the Soul*. The second book contains significant material on the *origin of our presentation of time*, including, in this respect, the thesis that motion and rest are counted among the common objects of sense (418 a 17). There is after all supposed to be no motion and no rest without time, and thus time appears as included in the *sensibilia communia*. It is contained in the objects of outer perception.

2. In the sixth chapter of the third book, Aristotle speaks of judgments *which relate to what is past and what is future* (430 b 1). 'If an assertion is of something that was or will be, then the time is added in thought and posited therewith.' Aristotle seems to have in mind that even these sentences express affirmative judgments and in fact categorial judgments with the time as predicate. Both

remarks are strange. How should 'something was' or 'something will be' be affirmative if the past is no longer and the future is not yet? And how should 'past' or 'future' be genuine predicates and thus apply to a subject that itself is no longer or is not yet? It is clear that in the passage in *De anima* III, chapter 6, according to which in every sentence in which one speaks of 'was' or 'will be' the time is added in thought, Aristotle cannot be speaking of time in the sense of a relation to the movement of the uppermost celestial spheres. Is he here taking time in a wider sense according to which it would be duplicated in every changeable object? Or is he not thinking at all of something that is counted or of something that could be counted, but only of something that is earlier or later than that which is present? The 'was' and the 'will be' are the simple expressions of past and future; they seem however to speak only of a distance from the present in opposing directions, and not of any measures of this distance.

In the same chapter it is said of the *connection of the predicate with the subject*, that this connection, i.e. predication, *is effected in a temporal boundary*. The thought of the subject precedes, that of the predicate follows on, and their connection is executed between these two times in the now, which — as end of the former and beginning of the latter — has a certain doubleness, analogous to a point in a line.

V. *Parva naturalia, De sensu et sensibili,*
De memoria et reminiscentia

1. In these smaller works further interesting remarks are to be found. *Parva naturalia* gives information about the origin of the *ideas of earlier and later* and about the only way these ideas can be clarified. If one then understands what earlier and later in motion and rest are, and what the expression number is supposed to identify, then the definition of time is clarified in all its parts.

2. From the work *De sensu et sensibili* we gather that for Aristotle time, which in *De anima* II seemed to be contained in the objects of outer sense, is apprehended by sensation also in inner perception. For he turns here against those who hold that there could be times which were too small to be perceived by the senses. Were this so, he holds, then we could not even perceive that we are. (Clearly, since the present is limited to only one moment and thus to a time quite without extension, and we are only in the present.)

123

3. In speaking here of a *sensation* of time, Aristotle departs from the doctrine in the fourth book of the *Physics* according to which time would not be if there were no intellective soul. This doctrine would imply that animals would enjoy no knowledge of time. They certainly do, according to *De sensu et sensibili* (Ch. 7, 448 a 19ff.), and also according to *De memoria*, where a sensitive apprehension of time is spoken of. This animals enjoy in so far as they have memory (449 b 29). They have a sensitive presentation even of that which is future, namely in expectation.

It is clear that a third meaning of time has here been added to the two already recognised (number of the first movement, and number of any sort of movement or rest). In this third meaning the moment of enumeration falls away. In this sense animals do not lack knowledge of time.

4. Since in memory one is conscious of one's own earlier experiences, the sensory object is here *psychical* (450 a 21). In contrast to this, time, as it is apprehended according to *De anima* II in motion and rest, is an object of the outer sense.

B. Criticism of the doctrine

Already in the presentation of the doctrine we had occasionally to insert critical remarks for the sake of better understanding. And indeed a good many would still have to be added, for however much of interest and importance is contained in the thoughts of Aristotle in relation to this subject too, there is much that is inadequate, indeed completely and entirely wrong.

1. Above all the question is not clearly put as to *which intuition the concept of time is derived from.* He says on one occasion that we experience something as earlier or as later *in sensation.* But how? It seems after all that sensation relates only to what is in the present. Thus one could suppose that memory has to be brought in aid. But how does that which is earlier show itself therein?

The appearance of succession in melody, speech, etc., is nowhere discussed.

The appearance of the present as two-fold boundary is characterised as such, but what is not investigated is the extent to which what is not present must thereby also make its appearance.

Nowhere is it said that the differences of past, present and future are the only ones that are given to us intuitively, and nowhere is there mentioned the variety of the *modi* of presentation that is called for here.

2. Aristotle says that only the present is, that past and future are not. But how the *historically factual* and that which is to come would be differentiated, on the one hand from other non-existents and on the other hand from each other, this is inquired about in no way in Aristotle's often subtle discussions, so that what is most essential remains unclarified.

3. He does not even stress that in order to think something as in some past or future *it is necessary for us to think something as in the present* and as at a certain temporal distance from the relevant non-existing something, but in a different way according to direction and magnitude of distance.

4. He does not dwell either on the fact that everything past or future that we apprehend is in the last instance *inferred from what is present*, albeit in different ways. For if we infer to something past, then we infer it as something whereby the present is directly or more or less indirectly determined. And if we infer to something future then as something in relation to which what is present has more or less directly exerted a determining influence. It is known in how close a relation many people want to bring the concept of cause to that of time, and indeed when past and future meet each other in the present it does seem that in this contact it is always only the past (or that which is ending) that has an effect on the future (or that which is beginning).

5. Very interesting but not completely harmless are the remarks about the *now*. This is supposed to be not a part of time but to bring about a connection of parts. It should certainly belong to a passage of time as boundary: for as he expressly points out, time without the now would not be, but without time the now would not be, either. But how the now is supposed to make a contribution to the length of time when it is itself without extension is not made clear; yet it seems to be agreed that it is only through the contributions of the successive instants that what is somehow to be acknowledged as the length of time can come to be. Aristotle sees in this difficulty an indication that time participates in the character of motion, which is for him an imperfect actuality.

Among his most interesting but still dubious remarks is his assertion that the now presents itself as end and beginning of something different and that time thus reveals itself as changing and as something that extends without beginning and end.

Interesting also is the aporia whether the now passes away into itself or into another and then in such a way that it passes necessarily into a now that is separated by a time containing infinitely many other nows. I do not find this aporia solved, nor either that according to which there appears to be no time, because neither past nor future *are*. Does he want to solve it by having the present belong to them both as boundary? This would be suggested by the remark above concerning the relation of now and time as conditioning each other mutually.

6. Also highly remarkable is the thesis that *if there were no soul, then there would also be no time*. One could be tempted to conclude that if there were no soul, then there could also be no motion, since to every motion there belongs time.

7. Aristotle also illegitimately denies that there could be a *first and last moment of a motion*. Further, that motions in contrary directions could follow immediately upon each other. Further, that in a moment taken for itself one could speak just as little of rest as of motion, that, however, in the case of a stretch of time during which something moves, motion would apply to every moment, just as, in a period of rest, rest would apply to every moment.

A rectilinear motion back and forth is for him not a unitary motion. Why not? Perhaps for this reason: because he is of the erroneous view that a rest would have to lie between the two parts? Circular motion is one and as long as it is not interrupted then it appears as such however often it is repeated. But what about motions along irregular curves? Is it here also the case that if the motion continues and no rest enters in then the motion is one? And how do things stand in the case of acceleration?

8. He accepts that one cannot speak of differences in *velocity in relation to time*, but fails to appreciate that time must still have a certain constant velocity which, when compared with the velocities of motions, is slower than the one, quicker than the other.

9. He touches on the difference between *primary and secondary continua*, but does not conceive and clarify it sufficiently.

10. He acknowledges that earlier and later are opposites and thus constitute a plurality, but it does not become clear *what kind of plurality* this is. He seems not to accept that it would be a substantial plurality.

11. Aristotle says that motion follows upon magnitude (and presumably means spatial extension, length of path) and that time follows upon motion. He seems also to see that the measure of time is the *measure of motion*, but it is not certain that he did not also see the measure of space as the measure of motion, which would lead

to the absurdity that in the case of motions of different speeds two unequal magnitudes would be equal to one and the same third.

12. One of the most dubious points of the doctrine lies in what it has to say about the *perseverance of something completely unchangeable*. Although something unchangeable could continue to exist during the course of a motion, he ascribes no temporal length to its continued existence. He seems not to conceive it as a duration, or at least not as a duration which would have parts. There is supposed to exist here no difference and plurality of what is actual, but rather *one* actuality — and no difference and plurality of what is possible either, for the unchangeable is pure actuality.

13. He is thus confronted with the task of making intelligible *how, from an absolutely unmovable first principle, one could explain the change that is in the world*. He has not noticed that the task is wrongly put. For it is not correct that the first principle must be thought of as absolutely unmovable. What is correct is only that it is at every time immediately necessary. Aristotle overlooks that fact that not all kinds of change lack immediate necessity and he has in consequence not seen that it is absurd to explain change from a first unchangeable principle.

14. The way in which he seeks to account for the *unity of time* seems also not to stand up to criticism. He sees this unity as lying essentially in the fact that every motion and every duration of what is movable exists at the same time as a part of the motion of the sphere of the fixed stars.

If one asks whether it would be in accordance with the concept of time thus conceived to designate past and future as times, then there seems to stand in the way of this the fact that we do not think of these in terms of astronomical motions, as we do when we say that something was in a certain year, or on a certain day, or at a certain hour. Admittedly here, too, we do speak of the past and of the future as a unity. And similarly when we say: lately, recently, once, a long time ago.

This could awaken doubts as to whether the unity of time, of which one says that all changes and all rest fall within it, would be a unity because the apparent motion of the uppermost celestial sphere provides the unitary measure for the determination of the order of everything that happens in regard to earlier and later.

One would think that the conception of time as a unity is connected with the fact that everything that happens runs in completely uniform succession through all degrees of futurity until it reaches being present, and then on through all degrees of being past, whereby every event has (and preserves) a determinate temporal relation to every other, whether as simultaneous with it or as separated from it as earlier or later in a certain degree. The distinction of past and future in time then comes to be made in regard to the instant which *is*, to which the past stands in the relation of being earlier, the future in that of being later.

One could also say that the unity of time rests on the fact that everything we present and accept, when we do not think it as existing, is thought with some temporal *modus* of past or future, all of which belong to one and the same continuous series of *modi* of thinking. Even what is thought as present can be thought only as presently ending or presently beginning or as presently proceeding or continuing to exist.

X. Our intuition of time is, like our intuition of space, indeterminate in relation to the absolute specific differences. It is only relatively specified

Dictated in 1916 *[T 3]*

The second half of this item was reproduced by Kastil from the original dictation in a shortened form; he then substituted two extracts from a dictation of 4 January 1915 *[T 17]*. [Editors' note.]

1. What are we to understand by *time*? It seems initially appropriate to bring forward examples of what one calls time. As such are to be counted an hour, a day, a year, a century, the one being called a shorter, the other a longer, time. Are these, now, differences of time of such a sort that all hours would signify times that were specifically the same? They seem, certainly, to be all of the same length; but one hour can be temporally very different from another, just as what is of equal spatial extension can be very different spatially in that the one is here, the other in a place one thousand miles away.

Do the expressions 'now', 'formerly', 'at one time', and somewhat more precisely 'most recently', 'directly', 'yesterday', 'last year', perhaps provide examples of what is temporally specifically different? These are answers not to the question 'How long?', but to the question 'When?', and it is the latter which seems to be most properly related to the peculiarity of time. There are also however other sorts of questions in regard to time which one cannot answer with a 'now' or 'yesterday' or 'last year'. 'What time is it?', 'What day of the month is it?' That what is, is now, is beyond doubt, and yet still we desire to know where we are in time.

If one investigates the matter more closely one finds that in the second case one is asking after an event that concerns the motions of the heavenly bodies. We order the events in regard to their earlier and later according to the axial rotation of the earth and according to its motion around the sun which we have recognised as virtually uniform. And thus, too, we incorporate into the total order of world events according to their earlier and later that boundary in the course of our life according to which it *is*. If such astronomical motions did not exist, then what now is would

129

nevertheless be, in regard to its time, just as it now is, and therefore no specific time determinations can be found in these relations to astronomical events.

Are such determinations perhaps given more adequately in determinations like 'now', 'formerly', 'at one time', or in the determinations present, past, future? If one says 'yesterday', 'tomorrow', 'a year ago', 'in a year', then on the one hand there is contained in these a determination as past or future, and on the other hand an attempt is made to determine more precisely, by appeal to the course of just those astronomical events, the degree which something past or future stands apart from what is present. Thus the determinations as past, present and future and as more or less remotely past or future *seem* to have the greatest claim to count as examples of specific determinations of time.[119]

2. If we look at what the philosophers have said in order to clarify the concept of time, then we find that Aristotle, where he wants to clarify his category of the ποτέ by means of examples, cites one or other of the determinations just mentioned, namely 'yesterday', 'a year ago' and he could of course just as well have chosen 'tomorrow' or 'in a year'. Perhaps also 'in the present' would have accorded with his intentions, though in the sixth chapter of the third book *On the Soul* he distinguishes among predicative judgments those which connect the subject to the predicate by the copula 'is' from those which bring about this connection by means of 'was' or 'will be'. He emphasises as a peculiarity of the latter that the time comes to be thought within them and is added to them. He thus seems to have believed that in this case we have to do with time determinations in a narrower sense.

Descartes defines time as duration and Locke follows him in this. Locke means thereby a fully uniform progression without beginning and without end in relation to which the duration of every finite and less regular progression would then be assigned its position in the world order as being earlier or later. Leibniz did not believe in such an absolute, regular course as something actual, and wanted to see in time not absolute but only relative determinations. It was for him the order of all things that happen in regard to their earlier and later. Here he stood in opposition to Newton, who believed in an absolute time which, conceived as eternity (as without beginning and end), was held by him to be an attribute of God. Every created thing would acquire a position in relation thereto, which would rest on an absolute peculiarity. Reid, too, who like Descartes and Locke had substituted for 'time' the word 'duration', believed that we possess an intuition of absolute time, though this is not given to us through perception but is rather suggested to us by the constitution of our human nature. When we reawaken in memory something that was experienced earlier, then we award it a place in this intuition of time. This intuition is certainly as such of a limited span, but it can through analogy be extended by us backwards and forwards into infinity. Kant, too, allows the intuition of time to be added as a consequence of our subjectivity to that which is revealed in inner sense, not however as one empirical datum among others, but rather as an a priori form available from the outset. Yet he conceives our intuition of time not, like Reid, as being finite, but rather as infinite. And while, according to Kant, time does not strictly flow, since it is the phenomena that flow through it, Schopenhauer once more speaks as is usual of a flux of time itself.

As one sees, none of these philosophers has been able to identify temporal differences other than those of present and past and more or less past, as well as future and more or less future, and so surely we too, if we want to find out about time, must keep to these differences.

3. But now it is above all unmistakable that if one says of something that it has been or that it will be, then in so far as it is thought of as past or as future it will not be thought of as something that *is* in the proper sense. To the extent that it is past it *was* but is no longer, and to the extent that it is future it *will* be but is not yet.

Further, whoever says of something that it has been or that it will be, while he certainly does not accept this thing itself, certainly does accept something else that is in the present: something that, in being present, stands apart as later than something that was and as earlier than something that will be, and indeed now to a smaller, now to a greater extent, even though we possibly do not specify precisely and with full determinateness what the magnitude of this distance is. Thus whenever we designate something as past or future we have to do with the acceptance of something belonging to that group of relatives which themselves are in the proper sense, while the terminus to which they relate *is not* in the proper sense. Thus for example in accepting a believer in ghosts, we do not properly accept the ghosts to which he is as such related, though we may then also say that we acknowledge them merely as believed by him. And so, too, in the present case. If we say of something that it was a

131

year ago, then we do not in the proper sense accept the event, we accept rather presently existing things as existing one year later than it, and then we may also say that we acknowledge the event as having been a year ago. When something is presented as past or as future it is therefore a matter of its being presented not *in modo recto* but *in modo obliquo*. And everything that holds in general of something presented *in modo obliquo* holds therefore of it, too.[120]

4. Yet still, like every other sort of *modus obliquus*, it does have its special peculiarities, and we must enter into a somewhat more detailed consideration thereof. We shall orient ourselves around certain examples where the relation to something in such a *modus obliquus* is especially clear. Suppose we hear a sequence of musical tones or of syllables and words in a speech. What is it that happens here? One tone is heard as present and then appears as past, initially as most recently past, then as lying somewhat further in the past, then as still more remotely past, and so on, until it fades for us entirely. In the light of what has just been said, this process is to be conceived as follows. From the first appearance to the disappearance of the tone there is always something that is being perceived by us as present: first of all the tone itself, then something that stands apart from it as later, and then something that stands apart from it by gradations as ever later and later. The tone itself is then no longer *perceived*, the word 'perception' being understood in its proper sense; rather, something standing apart from it as later is perceived as such. And we could then say in this regard that the tone is now perceived by us not as in the present, but as having been in the recent past. This is uniformly valid, whenever one tone follows upon another or the same tone is renewed constantly or also when a pause occurs. The presentation of the tone *in modo obliquo* will always include within itself something as present by way of a relation of later to earlier. The same explanation suggests itself for the hearing of a speech or the seeing of a motion.

5. But one riddle remains to be solved. How can something be compared with something and be found in this comparison to be different from it, if it is not itself presented, in some specific determination which pertains to it as such, along with that from which it would differ? If this is impossible, then the question arises, when we say one thing is later or earlier than another, as to what the determinations are in virtue of which the things or phases of a process lying temporally apart are distinguished from each other. In which respect does the note c sounding for the third time stand

apart as later from its materially completely similar predecessors, and does so by twice as much from the one as from the other?

If one says that the later stands apart in time from what was earlier and the earlier from what will be later, and this in exactly the opposite direction, then one says practically no more than this: what stands apart from something as later or earlier stands apart as something temporal which is, from something temporal which is not, and this in one of two opposing directions and in greater or lesser degree. It seems however that properly speaking nothing has been said of, for example, the third repeated c, which would not apply equally to its two predecessors, other than that the one *is*, the others are not. And then it appears hard to understand how the present c could stand apart from one of the two preceding c's half as much as from the other, for there is after all no intermediary between being and non-being. Indeed even the standing apart in opposing directions appears absurd, if we have to do here only with the difference of being and non-being.

Clearly we are referred here to the existence of temporal differences which do not coincide with those of present, past and future. Being past does not give to the tone that we heard earlier and is now past any new property, and being present can certainly not be seen as a property; at least not as one that would distinguish what has it from other existing things. 'Being present' and 'being' mean the same, and thus one can also equally well say of everything that is past that it is presently past.

6. What these temporal determinations are, on which are founded the relations of standing apart temporally to different degrees, should become clearer through the following considerations.[121]

We speak of rest and motion, of alteration and perseverance, of changing and unchanging continuance, of beginning, ending, going on. By means of the expressions 'rest', 'perseverance', 'unchanged persistence', 'changeless duration', we seek to exclude differences which are given in the case of the before and after of motion, alteration, beginning and ending. But not *every* difference can be excluded thereby, and this is already hinted at linguistically in expressions like 'going on', 'carrying on', where the 'on' signifies a certain departure from the previous state. This idea can be generalised. Like motion and continuous change, rest and continuous perseverance or duration, too, have a length, i.e. a magnitude, whose more and less is not thinkable without a plurality of parts, and this in turn not without some difference of parts. Even

133

in the case of an unviolated temporal perseverance, therefore, there must take place a real alteration, whether this can in itself be given to us intuitively or whether it can come to consciousness only in reflection upon other changes which we employ as measure for the length of the duration in question.

7. If these temporal differences were given to us directly in intuition, then we could explain without further ado how we are able to grasp things which in other respects do not differ materially as set apart in time. These differences are however in truth not given to us intuitively, as becomes clear when one considers the following:

a) When we fall asleep and wake up again after a long period, we are not able to give the distance of the now present moment from that moment which preceded our falling asleep.

b) We know that everything that is concurs in respect of its temporal peculiarity and must without exception differ in like measure from what lies apart from it in equal degree as earlier or later. We know, therefore, that in this genus of determinations different specific differences are incompatible, not only in the same subject but even in different subjects.[122] And we know, too, that while certainly other sorts of determinations can renew themselves in the same subject, the same specific temporal determination can never recur. It cannot remain — and it cannot reappear, either. Thus for everything that is, belonging to a certain species of this genus is absolutely necessary and belonging to any other species of this genus absolutely impossible. It would be absurd if something were now which did not have this specific peculiarity. But if we were to present this peculiarity in actual intuition, then this necessity would have to be capable of being grasped in this presentation, and were we to present intuitively another species of this genus, then it would have to reveal itself as absurd. As experience shows, however, we are conscious neither of such necessity nor of such absurdity in regard to any of the determinations given to us in intuition.[123] Thus there is left for us only to say that, however certain it is that the passage of time involves as such a certain continuous change, this change is transcendent as far as we are concerned.

c) Assuming, however, that something unintuitive to us were given to us in intuition, what would this apprehension as necessary or as impossible have to be like? The apprehension as impossible would have to diverge completely from the way in which we otherwise apprehend something as impossible. For if such determinations were revealed to us, then one and the same one, though completely the same in appearance, would have to show itself now as necessary, now as absurd. Of course it would not show itself as necessary in the sense of that which is necessary of itself. Since, as experience shows, no thing is necessary of itself, so also no thing is such that its time-determination would apply to it as something necessary of itself. For if it were not, then it would not have this determination either. This determination would certainly however, if it were given to us in intuition, refer us to an efficient principle of the relevant temporal thing, and from the peculiarity of this efficient principle the temporal determination of what it has produced could then be explained as necessary consequence. The efficient principle would clearly have to be its first efficient principle and indeed the first efficient principle for all other things existing at the same time. The principle itself however would have to be a moment of a necessary, uniformly proceeding change having as its consequence a continuous change of all the things conditioned by it. Such would be given if the first principle were to be thought of as all-knowing, all-willing and such as to determine everything through its power. For its knowledge would change as does the truth and its will would operate and take its effect with constant temporal displacement.[124]

8. Let us return to the question raised in 5. above. We have seen that, between a note c presented by its hearer as present and every note presented as past and as in greater or lesser measure past, there obtains not only *this* opposition: that the one appearing as in the present is thought of as existing, the other as non-existing. It seems rather that, while the tones enjoy the relative temporal determination of being in the present, there belongs to each one of them also another such determination, indeed perhaps one that stands in a precisely specifiable relation of magnitude of distance.[125] Our question is thereby however reduced to the question whether we can present temporal distances at all, and in particular whether we can present their differences of magnitude, in spite of the transcendence of the absolute specific time-determinations which underlie them.

One might be tempted to deny this, especially if one sticks only to the fact that distances are never given to us intuitively in the realm of sensory qualities except when the corresponding absolute specific differences are given too. Thus for example no differences of pitch are given without the presentation of certain absolute tones; no differences of colour without the presentation of colour-species. It would however be unjustified to generalise this requirement. This will be understood immediately when one sees that there are also many other relations of magnitude which are given without our bringing to awareness the absolute magnitudes between which they obtain. If, for example, someone has informed me that one man is twice as rich as another, then nothing at all is said thereby as to the absolute magnitudes of the fortunes of either. This example relates only to conceptual knowledge, but we need only turn our attention to the spatial sphere to see that differences can be given even in genuine sensory intuition without corresponding spatial positions. When we look out, then, if no earlier experiences provide us with additional points of reference, we apprehend nothing as regards the distance of the objects from us.[126] The location of what is seen appears to us in its dimension of depth without any absolute specific determinations. But does it then appear to us as absolutely determined at least according to length and breadth? Many have asserted this, and seen it as sufficient for this purpose that everything appears to us in the dimension of depth certainly without absolute determination but still with a certain relative determination, namely as entirely the same. But does this not prove exactly the opposite? For in relation to breadth and height there would apply quite different specific differences of places as well as quite different distances of the individual boundaries from each other. The object therefore appears to us in regard to the where of every individual point as differentiated through no absolute determinations of place, but only as differentiated through relative specific determinations. Every individual point is specified for itself through nothing; only when it becomes related to another point does every point appear as standing apart from every other in ever different directions. The individual point enjoys only the determination of being a spatial point, which involves its being a point in a three-dimensional continuum which, if this point is in full plerosis, allows the differentiation of a manifold of directions which is eight times as great as that of the angle of a cube taken as vertex of its stereometric angle. In comparing different points with each other we grasp the one as standing apart from the other and the one and the other as standing apart from a third to a greater or lesser extent, for example twice or half as far, or in a different direction. But all these differences testify to nothing concerning the absolute specific local differences of spatial points.[127]

The case is now similar in regard to time. In grasping something as temporal we apprehend it, absolutely speaking, only in regard to its general determination as temporal point which is as such a boundary of a one-dimensional continuum and — when in full plerosis — a boundary in two opposing directions of past and future. But this is nothing that would not apply to every other point in time, too, if it exists. This implies that it stands apart in indefinitely manifold fashion from indefinitely many other temporal boundaries of an indefinitely small continuum. These are however traits which every point in time exhibits in the same way.

Inner perception, which apprehends its object with evidence, is restricted completely to a single now. It reveals this to us as a boundary to whose nature there belongs that it bounds a continuum. Thus in spite of this restriction to the now we are directed intuitively in inner perception not merely to a single point *in recto* but also to a stretch of time *in obliquo*, though this is, in regard to its length, without any closer determination. In outer perception, in contrast, we do not merely present a single temporal point *in recto* with an associated indefinitely small continuum *in obliquo*. We present also other points, since we present that point which is given *in recto* certainly without any absolute specific determinations but still as standing apart in definite measures from others.

9. If both the existence and the transcendence of absolute specific time-differences is recognised, then this provides us with the key to solving difficult problems which have left many thinkers perplexed and in dispair. Thus the truth and existence of inner perception can now be perfectly well brought into harmony with the fact that everything that appears as in the present reveals itself without any temporal differentiation, every present as such as fully equal to every earlier present, even though a continual change of absolute temporal determinations is as certain as the fact that to be temporal is to have an end and a beginning. The view of the philosophers which would admit only relative temporal differences also becomes intelligible. Likewise the identification of time with duration and its exclusion from the *sensibilia communia*.[128] Quite special assistance is rendered however to the metaphysician who is disturbed by the question why the one temporal location of the world should enjoy a privilege in relation to every other.[129]

137

Part Three

SPACE AND TIME

I. Nativistic, empiricist and anoetistic theories of our presentation of space[130]

1906 *[TS 2]*

1. Where does our presentation of space come from? How has it come about that this question is considered seriously by science at all, for even the common man believes himself able to answer it without hesitation? I see what is spatial, I touch it, indeed I distinguish what occupies a place through every sense: even hearing allows me to differentiate easily between the buzzing of the left and right ears.

Thus the psychology of antiquity (Aristotle) knew no scruple here. It saw itself as having gained the presentation of space from sensation and saw this presentation as existing in every sensation. For sensation reveals to us not pure qualities but qualitative, spatially extended, shaped concreta, moving and at rest, and thus it clearly shows us various moments which involve the presentation of space and enable its abstraction. How therefore could it happen that doubts could have arisen here which in the end, as I am convinced, prove thoroughly unjustified?

It was not only the middle ages, ruled by Aristotle's authority, which stayed faithful to the conception of the common man. Even Locke in his famous *Essay concerning Human Understanding* shows himself to be at one with Aristotle in his results in this regard. He counts spatial extension among those simple ideas which, in contrast to the sensory qualities, are imparted to us through several of the senses. He is unmistakably thinking here not of a pure sensation of space but of sensations of concreta in which we find spatial as well as qualitative determinations. For how otherwise could he have contested the existence of innate ideas, since human beings do not after all come into the world without sensations of touch and vision (and even with closed eyes we still see the darkness

138

of the visual field). Yet in these sensations everything was concrete, and therefore unlike what Locke had called simple ideas, from out of which he then allowed the complex ideas to grow.

Yet Locke did not at all conceal thereby that the space-determinations of the real spatial world, in whose existence he did not doubt, do not always coincide with those given in our sensory images. This is after all known to the common man when he sees a stick as bent in water or when he says that a certain body appears smaller than another because the one is nearer while the other is further away. And Aristotle had expressly emphasised that the senses, which he held to be reliable in normal circumstances as to the 'objects peculiar to a given sense' are unreliable both with regard to 'accidental objects of sense' (as when someone who has salt in front of him supposes that he sees sugar, that is, something sweet), as also in multiple ways with regard to the 'common objects of sense' (that is motion, rest, magnitude, shape) — indeed the latter are most frequently subject to illusions.

2. It was therefore only a small step which was taken by Berkeley when he went on to make the assertion that, even presupposing the reality of a spatially extended world, none of the senses (and in particular not the sense of vision) is able to show us in even a single case the true spatial relationships, and that therefore it could only be something like experience which informs us of these. As Locke already saw, only on the basis of experience can we specify which spatial touch-sensations will correspond to a certain visual sensation of space: how many steps away is this, how many steps away is that seen object? And only on this basis can we establish for example that a ring that seems oval is in fact a circular band and that in another position it would in fact present itself as such to the eye. Berkeley therefore goes on to say that the presentations which we form for ourselves of the spatial relationships in the real external world (or rather in that world in which the non-idealist, in contrast to himself, erroneously believes) are given to us not immediately, through sensory impressions, but only on the basis of experience.

That he is correct in this is indisputable, and if the so-called *empiricists* had not wanted to go further than what was taught by Berkeley here in regard to our presentations of space then we should have to join them. However they do of course go further: they hold that in sense-impressions and especially in those of the eye, no presentations of space are given at all that are in agreement with each other and with the features of reality.

Adherents of this view in England ascribe it to Berkeley himself.

Thus J. S. Mill, an excellent representative of extreme empiricism, calls himself and his fellows 'Berkeleyans'. This is a misunderstanding. Berkeley was so much convinced of the spatially concrete character of presentations of sense, that he held a presentation of colour that would be freed therefrom to be completely impossible, just as he denied quite generally the abstraction of universals. But a doctrine is of influence as it is understood, and it is indeed possible that the whole of English empiricism in the realm of the theory of spatial presentations has taken its origin from this misunderstanding of Berkeley.

3. Yet it received further support from the fact that, because the metric space-relationships presented in sense-impressions are not the same as those in the external world, we do not at all measure the former when measuring the latter. In regard to the phenomenon itself, the means are not to hand which we use generally when measuring external things. In particular, there is no fixed, transferable standard of measure. Thus the length of a section of skin is not the length that is presented by the sensation of touch. We can perhaps bring sense impressions we receive through one eye into phenomenal coincidence with those we receive through the other, but never a vertical line with a horizontal. We have to rely on estimations or on the (by no means unobjectionable) method of enumerating just perceptible differences. Moreover, one proceeds with such estimations in a very superficial manner, and indeed usually omits them, since interest is normally directed exclusively to the spatial relationships of the actual world, not to those in our sensory presentations as such. The idea of the former, which has become firmly associated through habit, imposes itself so spontaneously that many allow themselves to be misled into taking it as something co-presented in the sense-impression. They do not notice at all that there exists something spatial that is different therefrom, something that, though it remains indeterminate in its metric relationships, no less includes relations of magnitude within itself. These are quite different from the ones we now wrongly suppose it to contain because we apprehend the latter on the basis of the former.

Hence the so-called Berkeleyans are completely right when they assert in regard to the presentations of spatial relationships which the common man ascribes to the sense-impressions themselves that they are much rather associated with these others and formed through habit. To prove this thoroughly was for them an easy matter. But then it seemed to them that, because the relations of magnitude which are actually presented in our sense-impressions remained almost without metric determinations, the presentation of what is spatial had been eliminated completely from impressions of sense.

4. Not only in England, however, but also in Germany there has for a time predominated an extreme empiricism which denies all presentation of what is spatial to our sense-impressions as such. And even today it has influential representatives. In Germany however the position has grown up not in connection with a misunderstanding of Berkeley but rather out of a very legitimate opposition to Kant's doctrine of an a priori pure intuition of space that is supposed, as subjective form, to become a receptacle for all phenomena of outer sense. If we imagine all such phenomena removed, then, according to this doctrine, space — infinite in length, breadth and depth — would still remain as something absolutely incapable of being thought away.

This is now something completely different from what had been taught by Aristotle in antiquity and in more recent times by Locke. What is spatial, like what is qualitative, was for them a matter of empirical facts, and Locke did not merely not believe in an infinite, pure intuition of space that would be given a priori, he did not even accept the possibility of thinking an infinite space except in negative fashion, for example in that one would think of boundaries pushed back and extended *in indefinitum*. Kant's psychological observations here were of no value, and he was also in the wrong when he made the possibility of geometrical reasoning dependent on the limitation of reasoning about plane continua of three dimensions. Mathematics since Riemann has ignored this limitation and has shown that topoids of arbitrary numbers of dimensions can be subjected to mathematical treatment.

But if the presentation of space is not a priori in the Kantian sense, must it then be given only through experiences, as the empiricists suppose? But why go over immediately to the opposite extreme, rather than return again to the Lockean doctrine that sensation provides us with concrete impressions that are both qualitatively and spatially determined?

5. The leader of the German empiricists in the theory of space was Hermann von Helmholtz.[131] Helmholtz was however influenced also by another philosopher, and this time positively. For Herbart, who held the continuum as such to be an inconsistent experiential concept and who wanted to reduce everything to a plurality of absolutely simple existents, could accept neither a

concept of the concretum with several marks nor any theory of the senses except one which allows our sensations to consist in a plurality of absolutely simple qualities. We are reminded of this when we hear Helmholtz confessing in his *Physiological Optics* that what sets him against the nativistic theory is above all the for him completely insurmountable difficulty of imagining how the stimulation of a single nerve could bring about a finished presentation of space without any prior experience. To be complicated and yet at the same time to be an immediately given datum appears to him as contradictory. But still, he adds, 'I acknowledge that this objection is perhaps too metaphysical to find a hearing in the sphere of natural science.'[132] He is therefore aware that such an argument will not make an equally strong impression on all. Hence we, too, need be not too shy in confessing that we ourselves belong to those left unconvinced. Other arguments would have to come forward if the empiricist theory of Helmholtz were to be made more acceptable.

Helmholtz does not fail to expound such arguments, and we find here essentially the same appeal to association and to the power of habit that we met already in the case of the English empiricists. The spontaneous power with which associated presentations press forth and the ease with which they appear and can be developed by us in whatever direction of thought are supposed to be what differentiatesintuition—i.e.intuitiveassociations—frommodesofpresentation of other sorts. Indeed he bluntly declares that as a result of the facility he has acquired in his investigations in regard to continua of more than three dimensions he is able to present even these in intuition.

The idea that presentations of space would be given already in our original sense-impressions themselves appears to him as a hypothesis that is not merely superfluous but in fact likely to bring difficulties of its own. For, as he puts it, such presentations would have to be overcome and suppressed through experience in order that the presentations appropriate to experience should assert themselves in their place.

It must be admitted however that he gets into no less significant difficulties when he considers the question as to what the presentations of space given through experience would actually be associated with. He supposes, with Lotze, that there have to exist local signs. Because these have no further interest for us in themselves, they are habitually overlooked, and indeed the tendency to overlook them becomes in the end so strong that it can no longer be overcome. Thus we are not in a position to specify what these local signs in themselves are. Only one thing can be inferred from the great ease with which those born blind are able to come to terms with spatial relationships, that they manifest in their sequence an order of differences analogous to that which obtains in the spatial things themselves.

Peculiar! If Helmholtz feels no compunction in declaring that the relations are as the spatial relationships themselves, but knows nothing of the absolute determinations, why does he not then admit also the hypothesis that these habitually overlooked moments of presentation in the sense-impressions would already themselves be spatial determinations? The reason he gives is completely insufficient. He holds that the spatial presentations given naturally in the sense-impression could not be sustained when those acquired through experiences of real space come to be attached thereto: they must either become transformed into the latter or cast out altogether. Elsewhere however he has himself asserted something quite different of the presentation of quality. In fact he goes much too far, especially in relation to the phenomena of simultaneous contrast, in that he allows a presentation of green, red, etc., to be so vividly associated according to circumstances with the same sustained grey, that the grey comes to be believed, through an illusion of judgment, to have been transformed into the green or the red. Nothing is therefore more natural than the reformulation I have suggested. This would however, as one easily sees, lead back from empiricism to the true Berkeleyan doctrine which is very close to that of Locke and Aristotle.

6. In the controversy between the so-called nativists and the so-called empiricists concerning the origin of our presentation of space it is therefore *nativism* that is without doubt to be given preference. Not however nativism in certain unscientific guises, where insufficient account is taken of the contribution of association, habit and experience both in our estimation of relations of magnitude in real space and in our establishing of equivalents of different spatial and qualitative sense-impressions as reference-points for the association of the corresponding determinations in real space; rather, nativism in that clarified and moderated form we find already propounded by Locke.

7. Yet there has recently arisen a new opponent of nativism that is by no means to be scoffed at. This holds nativism to be just as little in the right as empiricism, because we possess no true presentation of space at all. This remark may perhaps seem strange

and to stand in contradiction with the fact that we are after all, in speaking of space, not speaking nonsense and that it is spatial concepts with which a so highly developed science as that of geometry is supposed to be concerned. But do we not speak in arithmetic of billions and trillions without our ever being in the position truly to present to ourselves such large amounts? And does not something similar occur in theology, which confesses itself unable to form any true presention of the God with which it deals? It asserts that God is properly unnamable for the reason that for the word to become a name there would have to accrue to it the corresponding concept. One could therefore, along these lines, believe oneself able to accept without contradiction the idea that we are completely lacking in any true presentation of space, and that, like the true presentation of God or indeed of very large numbers, it is replaced by certain surrogate presentations. No one has, to be sure, declared himself explicitly in this way as regards the presentation of space, but the views of many respected thinkers do, when considered precisely, lead to nothing other than this. In fact we can count among them all those who conceive the concept of space as marked by contradiction — and that is not a few; consider, of late, Renouvier and his school in France, and for example Boltzmann among the German physicists — at least then, when they are of the opinion in regard to what is contradictory that just as it could not exist in reality so also not in an intuitive and unified presentation.

Although Ernst Mach deems himself to be a special sort of nativist, I do not in fact see him as belonging to the nativistic persuasion. Certainly he speaks of 'sensations of space'; but what he calls 'physiological space' does not correspond to the true concept of space for the latter requires continuity in three dimensions where Mach says of his physiological space that it is not continuous but is rather compounded out of a finite number of qualities. Now however the concept of physical space, too, is supposed to be developed out of this physiological space, in that, in place of absolute determinations there are substituted merely relative differences, a mere side-by-sideness which is in fact similar to that which obtains within the former. Thus the Machian physical space, too, is lacking in continuity, so that it does not correspond to the presentation that is to be connected with the word 'space' as this is employed within geometry.

It is clear that, according to Mach, a three-dimensional spatial continuum, as this belongs essentially to real space, would not only not be given in any single sensation; it would not be given either in many taken together, and nor would it be given in the presentations to be associated therewith (which are in any case supposed to be only weaker repetitions of sensations). All of them together could not possibly yield a concrete appearance that would truly correspond to what we understand by space. And how, then, is a general concept of space supposed to be capable of being abstracted therefrom? Such an abstraction is in any event impossible (and more especially according to Mach, who believes in no universal presentations at all), so that the alleged presentation of space would become a chimera.[133]

That no sensation (and thus of course also no reproduction of the same) would manifest a true continuum, whether taken singly or linked together, is especially emphasised by Poincaré. The illusion that it does, he holds, disappears immediately when one considers that no one has ever executed to infinity a division into parts. One easily sees that, on the contrary, there are already certain large numbers that one would never reach through such division. But without continuity there is no true spatial magnitude. Therefore a space is never present in our intuition. And Poincaré thereupon sets about explaining to us the way in which the concept of continuity, just as we have it, is built up. He begins with the whole numbers, continues by inserting between these the rational fractions, and arrives thereby at a series of numbers that is somehow 'everywhere dense'. Nowhere is there an actual hole, for between every two members, however close, there can always be found a third. We have then a continuity of the first order. In similar fashion he now inserts the irrational numbers and arrives at a continuum of second order. There follows then the insertion of the transcendental numbers, which yields a continuum of third order. And so, according to Poincaré, new infinite sets of numbers could continue to be inserted to infinity and continua of ever higher orders be constructed.

Only such intellectual labour, he assures us, and not sensory intuitions, could lead to the presentation of continuity. We must find in it therefore also a preliminary for our presentation of space. But how? Does one or other of the continuities here cited by Poincaré correspond to what the geometer demands that we conceive by space? Certainly not! For in each case there must still be missing points, where in space there would be nowhere a place without a point. And how could these gaps be filled in? Clearly only somehow through negation, as in the case of the idea of infinity according to Locke and of the idea of God according to the theologians in their doctrine of the *theologia negativa*. For we should have to say that if someone proceeding to infinity in his insertion of sets of points after the manner of Poincaré were to comprehend them all, then he would have arrived in his thoughts at that continuum which, bound up with certain other features, especially that of three-dimensionality, would correspond to the true idea of space. And thus we see that what was said earlier has been confirmed, namely that according to Poincaré we lack the true positive presentation of space and have available to us only a negative surrogate-presentation e.g. of something that can be unitarily divided in each and every way.

It would follow from this that nativism and empiricism were in like manner in the wrong, or the former still more so than the latter.

8. Yet such is not the case. Much rather is such a substitution of infinite point-manifolds for what we call the continuum to be completely rejected. Already Stallo, undisturbed by the authority of Riemann, protested against it in his famous book, *The Concepts and Theories of Modern Physics*. An intuition of a line, for example, can never be built up out of what is without extension. Were this the case, it would follow — to point to just one of the absurd consequences — that a smaller line could be brought into coincidence with a larger, since as Cantor has shown and as all mathematicians now teach, the points of a smaller line can be set into mutual one-one correspondence with those of a larger line, those of one half line, for example, with those of the whole. It is clear that, if the line consisted in its points, then the lines themselves would be brought into coincidence through the bringing of their points into coincidence piece by piece, which is impossible. And also the following is to be considered: geometry teaches that a line that is halved is halved *in* a single point. The line a   b   c therefore in the point b. And further, that one is able to lay the one half over the other, for example in such a way that cb would come down on ba, the point c coinciding with the point b, the other end coinciding with the point a. According to the doctrine here considered, in contrast, the divisions of the line would not occur *in* points, but in some absurd way behind a point and before all others of which however none would stand closest to the cut. One of the two lines into which the line would be split upon division would therefore have an end point, but the other no beginning point. This inference has been quite correctly drawn by Bolzano,[134] who was led thereby to his monstrous doctrine that there would exist bodies with and

146

without surfaces, the one class containing just so many as the other, because contact would be possible only between a body with a surface and another without. He ought, rather, to have had his attention drawn by such consequences to the fact that the whole conception of the line and of other continua as sets of points runs counter to the concept of contact and thereby abolishes precisely what makes up the essence of the continuum. It is therefore not possible that the concept of contact, as we all possess it, can be won from the formations manufactured by Poincaré. Indeed we would certainly not possess it at all, were it not given to us in some phenomenon or other through sensation or reflection.

9. It is a great mistake to believe that a contact of this sort given in sensation could not in fact be noticed by us since it would presuppose an infinitely minute differentiation of every part, however small. Consider a chess board with red and blue squares. If we were to split up each square into 64 smaller squares, once more alternately red and blue, and if we were to go on repeating this division arbitrarily often, then we would eventually come to a chequered field of red and blue squares so small that we should no longer be able to notice any of them individually. We would however still continue to perceive the total field, and not only this, but also its participation in red and blue. It would appear to us as violet, i.e. as reddish blue. And if half the squares were black instead of blue, then it would appear to us as reddish brown, i.e. dark red. And if there did not obtain for the visual sense that law according to which unilluminated parts must be filled in by black, then we should experience the red not as darkened but as a pure but less intense red, analogous to a quiet tone. We do not, after all, notice the individual particles of red when we perceive a surface filled intensively with pure red, yet we do notice that this surface is red.

Something similar holds now of the perception of contact. We do not notice that this obtains at this or that smallest location, since we do not notice the smallest location itself. We do however perceive that it is given in the whole that we see, not merely once or in some small quantum, but all in all in quite considerable quantity, and therefore that it is truly given and makes up precisely the peculiar distinctive character of the continuum.

Thus the objection is refuted completely. Were it correct, then it would count not merely against the possibility of a perception of continuity but also against the perception of a redness or quite generally of any kind of sensory quality.

We see hereby the last difficulty removed. Of the three mutually contesting views — of which we might call the last presented an *anoetistic* theory, referring to the two others with the traditional names of empiricist and nativist — the first is completely false. The second is justified only in a small part, in so far as the presentations of the spatial relationships in the real corporeal world which we form on the basis of our sense-perceptions are different from the spatial presentations contained in our sense-impressions themselves and are acquired through experience. The third, however, once clarified by this concession to empiricism, is to be seen as the correct and certain theory.

10. This much only can be conceded to empiricism, that with the spatial determinations given in sensation there come to be associated spatial determinations different therefrom. And indeed not such determinations alone: there are also qualitative differences (colour perspective, decrease in the strength of sound at a distance), which contribute in many ways thereto. These associated spatial determinations turn up as a consequence of habit and have as it were an instinctive natural force in virtue of which they become decisive for our estimation of the spatial relationships in the real world. Together with these there occur presentations of very different content, for example of movements or of the muscular sensations which we enjoy thereby, all of which should not however tempt us into confusing the presentation of space itself either with such presentations or with a mixture thereof produced by some sort of imaginary psychic chemistry.

11. Some nativists have called into doubt whether the local differences of different senses would be homogeneous with each other. A mere analogy would after all suffice. (Thus for example Stumpf.) Since in physics and geometry we have to do only with relative, not with absolute magnitudes and their relations, this could indeed be accepted without prejudice. Yet still, the doubt is not justified. In relation to taste, smell, warmth and touch, commonly held to be different senses, it would be absurd to deny the homogeneity of local differences, for in many particular cases they are specifically the same. These qualities suppress and combine with each other as colour suppresses and mixes itself with colour. (Cold food, rough and smooth food, for example diced or sliced potatoes, bitingly sharp smells, etc.) But we are able to supply the proof also for the local differences of the senses of colour and sound in relation to each other and to the remaining senses. One closes one's eyes and stimulates a feeling of cold in the nose by

breathing in cold air. The visual phenomenon appears then to be above the phenomenon of coldness. One closes one's eyes and stimulates a buzzing, first of all through pressure on the left earlobe, then through pressure on the right. The auditory phenomenon appears in the first place more to the left of the visual phenomenon, in the second place more to the right. That it will not do to put this down to our knowledge of the anatomical relations of the nose and ears to the eyes should be clear from the fact that one is not at all tempted to award the visual phenomenon itself a dual location in virtue of the separate location of the two eyes. The empiricists, too, know only one genus of localisation; if there were more than one for those differences which are the source for the differences added through association, then it would be to be expected that the latter, too, would be of different genera.[135]

12. I do not want to leave unmentioned the fact that, of the three competing views, it is the nativist conception which comes closest to that of the common man. Indeed we could express ourselves in a way similar to Laplace when he said that the calculus of probabilities is the vindication of healthy common sense. And in the doctrine of the origin of spatial presentation, too, the judgment of healthy common sense has shown itself to be in essence in the right. Just as the scepticism of Hume could not be the last word in regard to inductive inference, so in regard to the origin of spatial presentation the last word could not be the empiricism of Helmholtz — and still less could it be the anoeticism of Poincaré and others. Even today, I think, the last word would have been said already in nativism's favour, if only the proponents of the different views would always take the trouble to take sufficient notice of the arguments of their opponents.

II. The impenetrability of bodies in space rests on the fact that spatial determinations are substantial and individuating

Dictated 7 February 1915 *[T 6]*

Kastil produced two transcriptions of this item, having made appreciable changes in the second and evidently later transcription, above all to section 4 and to the notes. It is this reworked version that is included here. [Editors' note]

1. The theory of space is the theory of bodies as existing in space, i.e. as localised and as extended. These are, without doubt, real determinations; for after all, what is not real does not exist in the proper sense. Another question is whether they are substantial determinations. Views are in this respect divided. Descartes saw in spatial extension the essence of corporeal substance. He apparently did not however see change of place as a substantial change, which contradicts his former assertion since extension, as sure as it is spatial, includes locality in its concept just as a length of time includes the concept of temporality. Still, he judges no less superficially when he sets the essence of mind in thinking without accepting change of thought as change of substance.

Others have, unlike Descartes, denied that extension is a substantial determination: some in that they counted it among the accidents of the body, others in that they interpreted it as something that would be neither substance nor accident but would exist for itself in such a way as to stretch out to infinity far beyond the reach of bodies, and form the presupposition for the latters' existence. This view was held by Locke and Newton. It is clearly untenable, for what else could a thing like the Lockean space be, other than a special sort of substance? The services it is supposed to perform are however quite superfluous. An empty space needs to be something positive just as little as does the absence of a tone when a tone is skipped in playing a scale. Among those who saw spatial extension as an accident of body there belongs Aristotle. Not he, but many of his indirect disciples then conceived extension as standing in addition in a relation of subsistence to other corporeal accidents, as especially to the sensory qualities, use of which was then made by the scholastic theologians in their doctrine of the eucharist. Considered from the point of view that it is demonstrably true that certain accidents serve as subject for other accidents, this would be nothing inconceivable. If however one asks why extension and place were regarded as non-substantial differences, then the reason is presumably to be found in the fact that changes of place leave the body, considered for itself, practically unchanged in its peculiarities. Only its proximity to or distance from other bodies is altered thereby. Thus change of place seems so to speak to be the least that a body incurs, whereas one generally understands by substantial change something peculiarly significant, having as consequence a far-reaching change in the determinations of whatever undergoes it.

Since something of this sort occurs in chemical transformation, one is tempted to hold this change as a substantial change in the body. Indeed J. S. Mill wanted to distinguish the concept of substance from that of accident precisely through the fact that there is attached to the former an inexhaustible manifold of specific peculiarities which can be detected through special observations. Yet this is not the true definition of substance if such is to secure for the name its original meaning, and in order to decide whether something is a substantial or an accidental determination quite different criteria are required. The concept of substance is not exhausted in its being the subject of something other, since, as already said, there are also accidents which are subjects of other accidents. Much rather is it a matter of being such that it itself has no further subject. But how can one be sure that this is so? I answer: substance may not fall away, without it being the case that another species of the same genus takes its place. Thus for example the thinker as such is not a substantial difference, since something capable of thinking could still exist even without actually thinking and yet still remain the very same thing.[136]

The verdict on the question whether place and extension belong to the substance of body depends, therefore, on whether the localised and extended thing could still be the same thing if it existed in no place at all and had no extension at all. No one will accept this, and thus we must clearly allow differences of place and therefore also of extension to count as substantial.

2. Nothing is as yet thereby determined however in regard to whether the differences of place of the body would be its only substantial differences. The opposite will already have been demonstrated if it has been shown that for every thing the determination of time belongs equally to the substantial differences.

And the question still remains open whether there perhaps exist still further differences, belonging to a third series, which would be substantial,[137] or whether we have reasons to rule this out. If we contemplate the space of our visual field, then we find that it is in fact qualitatively filled in all its parts, for if no light would fill it (to which we have to count besides white also colours in the narrower sense such as red, yellow, blue and their mixtures), then it would be filled by black, and this is not somehow to be identified as the absence of all qualities but is a quality no less positive than white. In other senses, however, in contrast to that of sight, it is often the case that the qualities fall away, sometimes partly, sometimes completely, and connected with this difference is another peculiarity, namely that all senses other than sight manifest now greater, now lesser intensity in what appears. The diminution of intensity is however to be attributed to nothing other than unnoticeably small qualitative gaps. If in some concrete case there are given no qualities at all to these senses, then there is also nothing of place and extension to be perceived. If, for example, a quality of smell had never arisen, then we should also have no idea of the field of smell. This could lead one to infer that space without quality cannot be a phenomenon at all. One might then suspect however that in analogy thereto there would have to belong to a corporeal substance in addition to its temporal and local differences, also a further difference, from a third series, which would be superadded to the determinations of place like the qualitative differences in sensory space. Without such an analogue of quality, one could say, one would have an empty space which would no longer be anything positive and real. This would harmonise well with the fact that a body lacking in all peculiarities other than those of temporality and location is nowhere to be encountered. Bodies are physically and chemically multifariously specified, and even if one pays heed to what all bodies have in common, there is much in them that does not allow itself to be found through analysis of place or extension. Thus for example all bodies have a certain elasticity and this is inconceivable without a certain special connection of parts and an active attraction and repulsion, where nothing of this seems to be contained in the presentation of place and extension.[138] All of this could speak in favour of a third substantial series, if not of a number of such. Admittedly we have no intuition of their specific differences. We do not, after all, have the right to think of the sensory qualities as belonging to the real world and nor do we have any presentation of something analogous thereto. This is however no great loss, since after all, even in the mental sphere, where we grasp the substance only as to its most general concept, we have nevertheless to be convinced of the existence of specific substantial differences.

3. Yet grounds can be brought forward also for the opposite view. Above all it is undeniable that the determination of place is something positive. And already for itself alone, so that even assuming there would somewhere exist something determined as to place without addition of a qualitative determination, we should still not be allowed to speak of an absolutely empty place. For one could speak of absolute emptiness only if there was in reality no possible location at all. The appeal to the fact that, in regard to our senses, if a locality does not appear as bound up with a quality, then it does not appear at all, is for this reason able to prove nothing.[139] Things are so as a matter of fact, but the opposite was conceivable at the outset, just as certainly as it can happen that we present to ourselves what is located and extended in abstraction from every quality and make this into the object of geometrical investigations. Further, it is not only Descartes who held on to the idea that the entire substance of body consists in the spatial extension that is positively unique to each. Locke, too, who fought against him, knew nothing better than to add 'solidity', i.e. impenetrability,[140] in order to complete the concept of a body as compared to that of a space that is not filled by a body. One notes that he held empty space, too, as something positive and real but not impenetrable. Certainly this idea of space seems strange. If it were something that allowed the distinguishing of different parts, none of which were impenetrable, then why should not one part, by being displaced, be able to penetrate another? It is here apparently presupposed that it is impossible for one part of space to penetrate another, and with this seems to be connected the immovability of each such part, just as in the case of bodies. In fact from a body's impenetrability by other bodies someone might infer their immovability, in case bodily things were given in a juxtaposition which has no gaps and no end.

4. To this must now be added another not unimportant aspect. One asks whether the impenetrability of bodies, one of the most well-established theorems of physics, is guaranteed for us only through experience or is already clear a priori. Some are perhaps inclined to decide in favour of the former; then, however, it would be a logical mistake to hold the impenetrability of bodies as established in its generality. An inference from the known cases to a new individual case may indeed be drawn even with infinite probability, but an inference to the exceptionless general law is not

permitted by such an *inductio per enumerationem simplicem*. Even the inference to two new cases would be less probable than that to one. And in regard to the inference to a plurality of new cases which was no smaller than that of the observed cases themselves, the probability for and against would be equally balanced, and if one then inferred to an unbounded plurality of new cases it would in fact turn out that in spite of our imposing *inductio per enumerationem simplicem* the probability of the law would be vanishingly small.

Things would be quite different in the case of an a priori evidence. But how should such an evidence be intelligible? How should there be a contradiction in the penetration of two bodies? I answer: in no way at all if there were, in addition to temporality and spatiality, also a third series of determinations which belonged to the nature of body and, combining with these others, co-determined the individuality of each substance. Certainly however if in these two series there is given everything that belongs to the individuality of the substance. For then it is guaranteed by the *principium identitatis indiscernibilium*. The assertion that two bodies had the same determination of place would then, after all, say nothing other than that they are two without being distinguished in any sort of determination, which was recognised as absurd already by Aristotle long before Leibniz had formulated his *principium indiscernibilium*. Imagine, in order better to understand the argument, two corporeal substances at different places, matching each other not only in their temporal determination but also in any other substantial series there might be (rather like two red patches in different places in the visual field). Why should the one not penetrate the other? Because there belongs to penetration the duality of the substances involved, but this duality would be immediately abolished if there should cease to be a difference of local determination. For they would then be completely without difference, and thus no longer two. The matter would be different if these two temporally equal but locally different substances were to differ also in regard to some other substantial series. For then place and time alone would not determine their individuality, but this third substantial determination would have a supplementary individuating effect, so that one could no longer say that to eliminate the difference of place would be to eliminate the duality also. There would then be no contradiction in their mutual penetration. However, experience shows that bodies with different determinations are no less impenetrable than those whose determinations are the same, and if this is so then impenetrability could no longer be derived from the *principium indiscernibilium* even were we to find in relation to some quality additional to place and time something that characterised this quality as substantial.

5. While, however, there are very significant differences in the physical and chemical properties of bodies, this still yields no objection to a conception of them as merely accidental differences. For the latter can be incomparably more important than their substantial differences, just as, appealing to a comparison with the mental, the actual thinking that is added as accident to a soul is incomparably more important than the substance of the soul itself. For it is in thinking that there are to be found the oppositions of knowledge and error, noble love and the preferring of what is bad, so that Aristotle could say that a mind without actual thinking, lost in eternal sleep, would be devoid of all dignity. Moreover, it can be shown without difficulty that all physical and chemical phenomena can be made equally intelligible on the hypothesis that they are a matter of mere accidental circumstances and transformations. Indeed this is in a certain sense to say too little, and one could point to certain recently established facts which the physicists have sought, so far in vain, to bring into harmony with each other, which however appear immediately to be compatible without contradiction on the basis of this new hypothesis.[141]

III. What we can learn about space and time from the conflicting errors of the philosophers

Dictated 23 February 1917 *[TS 7]*

1. Our language possesses the two expressions *place* and *space*, corresponding to the Latin *locus* and *spatium*. The Frenchman says *lieu* and *espace*, but he has also the words *endroit* and *place*, similar to the German *Stelle* and *Platz*.

Time is contrasted with space, but also the word *time* with the word *place*. The Greeks contrasted *topos* with *chronos* and I do not know whether they employed also another term in addition to *topos*, as scholars using Latin had besides *locus* also *spatium*. Did they use *topos* in a twofold sense, or do *place* and *space* mean the same? And if the former, then does the question 'where?' relate to place or space? And is also the analogous expression *time* not unambiguous? And does the question 'when?' relate to time only in *one* of the meanings attached to this name?

2. One speaks sometimes of space as of something single and unique, sometimes also of spaces. As for place, it seems that only the latter is usual, except for someone who would imagine that, similar to other general concepts, the concept place would exist also as a universal.[142] Is place perhaps synonymous with space when one speaks of a plurality of spaces? Or is there yet another difference, in so far as one speaks of place even in regard to a mere point, not however of space, since one understands by the latter always the place of something extended, indeed extended in length, breadth and depth? One speaks of large spaces, of amounts of space, not however of large places and amounts of place. The fact that one tends to say in German *an einem Ort* [at a location] and *in einem Raum* [in a space] seems to be of slight significance since the French say *dans un lieu* and the Latins *in loco*. The term here refers to what is at a place and in a space, not however to what is in space as conceived in unitary fashion. Something similar holds also of *now* in relation to times and points in time in contrast to time as unitary. And thus also of *then*, to which there corresponds the *when*?, as *where*? corresponds to *there*.

3. What does one mean when one speaks of space and of time as of something unitary? It seems that here, too, different concepts are involved. Thus time (but not also space) was made into a mythological figure, into the God *Chronos*. One imputed to it an

activity, a generating and consuming of its children. Others understood by unitary time a completely uniformly proceeding motion without beginning and end in so far as this would allow us to determine numerically the relations of earlier and later. The motion of a highest celestial sphere, believed to be in eternally uniform rotation, was supposed to be time in so far as this rotation provided the measure of earlier and later. Others again held on to the idea of an eternally uniform change as the standard for a numerical determination of earlier and later, but did not believe that any such change was given in actuality. Hence they denied any actual unitary time and supposed that such could exist only in the improper sense, namely in our mind, where we are supposed to have artificially constructed it. Yet others were not in agreement with this, even though they too did not believe that such a unitary time would actually exist or (as they said) exist in itself. Thus Kant, and others who reveal themselves as influenced by him,1* held much rather that it was a form of intuition given to our inner sense from the very start. Whatever we present intuitively with the inner sense, would be presented as existing in this unitary time. Yet others believed in the existence of such a unitary time without beginning and end not as something merely phenomenal but as existing of itself, and believed that its necessary existence is immediately evident to us from the outset. If however one were to ask what this time is, then the accounts would diverge. The one group would say it is not a thing but a precondition or presupposition of all things, something that accommodates them as a sort of container. Thus Reid, and in recent times Marty. The nature of this condition, they say, is not to be further explained; it is something wholly quite peculiar, grasped by us immediately in its peculiarity. God, too, is conditioned thereby. Others, on the other hand, wanted to understand unitary time so conceived as

---

1 *It happened very often and still happens today that memory is counted as part of the inner sense, which is utterly wrong, since memory is not a perception, neither inner nor outer. Moreover, it does not have the evidence of inner perception. Not what we remember is perceived, but our remembering, i.e. only that peculiar belief that it is something past and that it was experienced by us. Kant presumably was guilty of the same confusion and so came to make time into an intuition of the inner sense. What is certain is that when we visually or aurally perceive a rest or motion, a continuous sound or a series of sounds in speech or melody, then what is coloured and what is sounding are given to us in external intuition with both spatial and temporal differences.[43]

something real, and since this real something is such as to present itself to us as immediately necessary they supposed that it must be an attribute of God, namely his eternity. Thus, explicitly, Clarke, but perhaps also Newton. If one wanted to ask in what relation this attribute stands to the essence of God, then the answer could presumably be no other than that it is not an accidental but an essential attribute of God, and thus one would be dealing here with the nature of God, i.e. with God himself, who would hereby perhaps be grasped in an indeterminate, non-exhaustive way. Clarke wanted to base on this a proof of the existence of God. The reference to the immediate and absolute necessity of time was supposed more particularly to refute Hume's assertion that no thing could be immediately necessary since the simple denial of a thing would involve no contradiction, which could obtain only between affirmation and denial.[144]

Opposed to all the mentioned conceptions were those Christian thinkers who taught, like Augustine, that time had had a beginning and was created by God in the creation of the world. God himself, as a completely unchangeable essence, has according to Augustine no temporal determinations. Hence one cannot speak of a length of the divine life and of a persistence of the same; for without difference there is no plurality and without plurality of parts there is no length. Time therefore began with the changeable world, and one cannot speak of earlier, empty times. Should we say that time is a unitary thing that was created like the other things with the creation of the world? Yet it seems inappropriate to ascribe such a view to Augustine. Perhaps he meant that the duration of the changeable world is the time of the world, which would be something comprehending within itself the duration of every single creature, something that would embrace them all within its unity. It would become therefore in some way an attribute of the world, just as for Clarke it was supposed to be an attribute of God. Yet much remains thereby unclear. Already we get no answer if we should ask whether, according to Augustine, time is a peculiar completely uniformly proceeding change in which everything changing in whatever respect would participate for as long as it exists. And it seems just as little possible to determine with certainty whether the earlier Christian thinkers saw the time which they conceived as having been initiated and created with the world as something absolute or merely relative.

We touch here upon an opposition in which the views of Newton, Clarke, Euler and many others diverge from those of

Leibniz and of many natural scientists of the present day. Leibniz declared time to be something merely relative. It is, according to him, nothing other than the order of succession of the things. Such a succession would be unthinkable without plurality of the succeeding parts and thus an individual for itself would be without all temporal determinations. The other party — however else they might differ in regard to the nature of time — affirms in opposition to this that every individual has already for itself an absolute temporal determination, a temporal position, in that it exists in an absolutely determined point of time absolutely conceived. Curiously, they also differ further among themselves in that some believe that every temporal point is, considered in and for itself, exactly similar to every other. Temporal points are, certainly, many; but here the proposition would not hold that plurality is given through peculiarities of marks which the units would have. The *principium indiscernibilium* would have here no validity. Others, however, will have nothing to do with such an assumption. Clarke can be counted among the former, while Marty belongs decidedly to the latter group.[145] One could however believe that a corresponding difference of opinion is also given among those who believe in no unitary time but only in many times, of which each coincides with the duration of the relevant temporally existing thing, according to whether they do or do not see in every moment of an enduring existence a special distinguishing mark.

4. Something quite analogous to what was said about time also holds in regard to space and to accounts of space. Those who speak of a unitary space are thinking sometimes of something that could serve as basis of reference for all determinations of place. Just as they wanted to have time understood as a complete, uniformly proceeding motion in relation to which everything could be determined numerically in regard to its earlier and later, so they understand by space something absolutely at rest — by preference a circumcluding boundary which is at rest. Thus according to Aristotle the outermost boundary of the corporeal world is supposed to be a spherical surface which is never displaced. He saw it as the *topos* in which the spatial bodies which are part of the world are to be found. Someone else could however perfectly well say that such a basis of reference could be provided by a quadruple of fixed points, in that from the specification of the distance of an arbitrary point from each of these, there can be derived a determination of its position in space. It has turned out that the *topos* of Aristotle does not in truth exist, and nor does anything in

the world of our experience speak for corporeal points in absolute rest. The fixed heavenly bodies do not themselves live up to their name.

Commonly however those who have spoken in recent times of a unitary space do not think of something corporeal in absolute rest as their basis of reference for numerical determinations of juxtaposition. They associate therewith a quite different idea, which shows a far-reaching analogy with what they understand by unitary time. And just as we saw in regard to the latter that opinions widely diverged, so this same comedy of opinions is offered again in regard to space. Some teach that, just as time is an intuition of inner sense given from the start, so space is such an intuition of the outer sense. Whatever we would present to ourselves intuitively with the outer sense, we present as existing in this unitary space. Thus Kant, and those who were later influenced by him. The unitary space they held to exist phenomenally in three dimensions, being unbounded in each and such that it could not at all be thought away. In itself however it does not exist. Itself immovable, it would underlie all phenomena of movement.

Others however believed in a space of this sort underlying everything corporeal not merely as something phenomenal but as existing in itself. And they held that its necessary existence is immediately evident to us from the very start. Thus for example Clarke and Reid. If however one asked what this space is, then opinions diverged. The one said it was not a thing but rather a precondition for all corporeal things, which would be contained therein as in a receptacle. Thus Reid, and this view is approached in recent times also by Marty. Marty, it is true, conceives space in other respects, more specifically in regard to infinity and immovability, in precisely the same way as does Reid, but differs from him in believing it to be created. It was reflection on the conceivability of topoids of arbitrarily large numbers of dimensions which led him to this deviation. Thus God, who is held by Marty to be conditioned by time, is supposed not to be conditioned by space. The nature of such conditioning, however, in the case of time as of space, is to be seen as something quite peculiar and not capable of being further described.[146] Others in contrast wanted to regard unitary space — which they considered to be a precondition for bodies and indeed for everything that belongs to the world — as something real; and because they belonged to those who believe that its existence is immediately evident as something necessary, they supposed that this space, too, must be an attribute of God, namely his immensity. Thus Clarke. Newton went so far as to describe space as the sensorium of God, an expression which seemed to Leibniz very offensive and whose glossing over caused much bother for Clarke. This attribute could of course be accidental just as little as could that of eternity, and thus here again one would have to do with God himself, who would however be grasped not quite completely. As in the case of the necessity of time, Clarke wants also to exploit the necessity of space for the proof of the existence of God. Reid has his doubts about following him in this; for although he takes space, like time, to be necessary not merely phenomenally but also in itself, space is not for him a thing, neither substance nor accident. The fact that according to Reid the visual sense reveals of spatial things only their extension in two dimensions, and that only the sense of touch leads us to the presentation and knowledge of the third dimension, does not seem to him to be a contradiction, since it points only to a more incomplete apprehension of space in the visual sense. This however leads him to the idea that even the three-dimensional presentation of space might still be incomplete in the sense that, taken in itself, it could possess even a fourth and, who knows, perhaps further dimensions, which would then (should a follower of Clarke concede this) lead also to expansion in the idea of God's immensity. Like Clarke, Reid too holds space to be non-created, but to be a non-created non-thing.

Just as, in relation to time, we found certain early Christian thinkers in opposition to other philosophers in virtue of having ascribed a beginning to unitary time, so also in relation to unitary space in virtue of their conceiving it as finite in all its dimensions. The space that is filled by bodies does not extend beyond a certain boundary, and just as there was no empty time before the time in which the world is, so also there are no empty spaces beyond the boundaries of the space that is occupied by bodies. Yet when it comes to the question what this space filled by bodies is, whether something for itself or something accidental to the corporeal substances (for example something yielded by the totality of their extensions), we receive no satisfactory explanations. Neither do we in regard to the question whether it is something absolute or relative, though the former is perhaps more likely.

As God has no time, so also he has no space; indeed even the world of spirits is not in space as is the world of bodies. We hereby touch upon an opposition in regard to space that is analogous to what we found in regard to time, in that many moderns, when they speak of a unitary space, believe themselves to refer to something absolute, others however to something merely relative. As already said, Newton held it to be something absolute that exists in itself; Kant held it to be something absolute that exists only phenomenally. Leibniz, in contrast, declared it to be nothing other than the order of bodies in regard to their juxtaposition. Already Newton however experienced the doubt as to whether it was not perhaps something merely relative, though he saw this as having been removed by the fact that when a sphere turns on its axis a centrifugal force makes itself noticeable. This would not be the case if the sphere were fixed and everything surrounding it were to turn in the opposite direction.

Euler, too, contradicted Leibniz in that he emphasised that it makes a difference whether one single body or the totality of all remaining bodies is set in motion, even if the one and the other motion would lead to the same relative positions. He connects this however with a further, peculiar doctrine which to a certain extent brings him close to the Leibnizian view. Leibniz had denied that any point would possess an absolute spatial determination; should several points exist, however, then in spite of the lack of absolute determinations and specifications for each single point, spatial differences and distances would exist between them. Euler, even though he believes in absolute determinations of place and thus also allowed a place even to some lone existing point, still held that the determinations of place, though many, are none the less such that none of them has any specific distinguishing mark. According to him, therefore, the *principium indiscernibilium* does not hold for the absolute determinations of place. In his polemical correspondence with Leibniz, Clarke had not wanted to concede that his *principium indiscernibilium* is plausible at all, since God's omnipotence could create arbitrary numbers of complete similars. Each would then be another, but would not differ or be distinguished from the rest in any mark. Here, too, it is to be noted that those who thought in this way about space thought analogously also about time. Each temporal point is for them indeed another, but without any distinguishing peculiarity. If it is in the present then it manifests itself exactly as do the others. The fact that it is an absolute point, and a different absolute point, seems therefore according to them not to imply that it has some absolute peculiarity which marks it out.

5. We could go into still greater detail in this survey of the views on space and time expressed by the philosophers. What has been

162

said will however be enough to give an idea of the disunity which prevails and which, on superficial consideration, could give rise to the foreboding that no certainty is to be gained here at all. This is however not so. On the contrary: it can be shown of very many of the views presented, however worthy the thinkers who have put them forward, that they stand in contradiction with clear facts or with themselves, and in revealing the mistakes as mistakes we shall have revealed also the path which alone leads to truth.[147]

Thus I say first of all that space and time as we speak of them when we say that all bodies are in space and that everything that happens happens in time are to be understood *not* as something corporeal at rest somewhere, whether it be a celestial sphere that never leaves its place or a number of fixed points or immovable coordinates which could serve as basis of reference for all other determinations of place. And similarly one is not allowed to think of an eternally uniformly proceeding motion, as Aristotle supposed there to be given in the rotation of the highest celestial sphere as basis of reference for the determination of the order of earlier and later. Such absolutely fixed bodies and absolutely uniform motions are not known to us in our experience.

I say further that the view is to be rejected that space as well as time are to be understood as individuals existing merely phenomenally and extended to infinity in every direction. Such a phenomenal existence would boil down to nothing other than the existence in itself of a haver of intuitions, and since the doctrine is supposed to hold of every one of us, it would imply nothing other than that there exists a large multiplicity of us who would have the same intuitions.

But what does it mean to say that these intuitions would supply a *form* for everything that is otherwise intuited by the members of this multiplicity, given the fact that these intuited things may be different for each and may change continuously in each in manifold ways? Even such differences and changes in the phenomenal sphere could be nothing other than differences and changes which do not truly and properly exist. What properly exists is a difference and change of the one who intuits.

Hence it will not do to search for what is temporal (and still less for what is spatial) in the merely phenomenal sphere. Much rather is it the case that if something is phenomenally in time, then it certainly is also *in itself*.

This was asserted before Kant by Thomas Reid, and indeed for time as well as space. But even though he believes both to exist in themselves in limitless extension, he still believes them to be nothing real; they are not themselves things, but only receptacles for things. Here, certainly, we have a case where concepts fail and a word comes patly in to serve their turn. If time and space are not things, how then could they fall under the concept of a receptacle, or indeed under any concept at all? The concept of a thing is after all the unitary concept for all objects of our thought without exception. Everyone who thinks — the word understood in Cartesian generality — thinks *something*, and if the concept of thinking is to be a unitary one, then so also must the concept of this something and we apply to it the name 'real', taken in its unitary meaning. Thus already Leibniz, too, when he touched upon the question of space and time in his correspondence with Clarke, hit the nail on the head with his remarks that the unitary infinite space and the unitary infinite time which Newton wanted to have underlying all other things would be none other than real subjects, real substances, which would have everything else as properties in themselves. Hence Reid's idea is an impossible fiction, and with his theory there is condemned also the theory of Marty which corresponds to it in all essentials. When Marty conceives infinite space as being, even though supposedly nothing real, still created by God,[148] he shows especially clearly that he does after all award to it the character of something real, even though denying it in the same breath.

Newton and Clarke, like Reid, conceived space and time as individual unities unbounded in every direction and existing of necessity. They, however, held it to be something real, and thereby avoided the absurdity that lies in the assumption of something that is and is yet not a thing. But what is this real something supposed to be which exists with immediate necessity? Newton and Clarke were led quite understandably to believe that they had to do here with attributes of the first principle of all things. But is it not a great presumption to suppose that such divine attributes would be given to us in intuition? And how is this supposed to happen? Are they objects of experience, like that which we apprehend in so-called outer and in inner perception?

Reid, who went more thoroughly into the matter of how we arrive at the presentation and knowledge of this unitary space and unitary time, denied this resolutely. Much rather does the idea of the one and of the other associate itself in peculiar fashion with what we perceive, whereby however space and time are not, like the objects of perception themselves, apprehended merely as existing, but rather as existing of necessity. But is this really more than the workings of imagination? As for me, I must definitely deny that the existence of infinite space and time imposes itself upon me as immediately necessary or that I have an intuition of the one or of the other. And that what holds of me corresponds also to the consciousness of others is seen in the fact that very eminent thinkers have denied the necessity and even the existence of a time without beginning and of a space extending infinitely in all directions. Thus Aristotle conceived space as bounded. Augustine propounded a beginning of the world, and Leibniz too held this to be a possibility. This is to accept also a beginning of time itself as possible, since for Leibniz the life of God is supposed to be without a temporal course and time nothing other than the order of succession in the world. And just as no time would, according to him, have existed without beginning, so also no space.

I need hardly bother to mention that David Hume, too, would have protested against the assertion that we cannot avoid the idea that time or space is immediately necessary. Such necessity would according to him signify nothing other than that a contradiction lay in the denial, but a denial could only contradict an assertion and such is here totally lacking. Kant for one part adopted this Humean idea, and if he none the less seems to ascribe to space and time a certain necessity, still it is clear when considered more closely that he does this not in the proper sense, since space and time do not after all, according to him, exist in themselves but only phenomenally.

Thus we have to do, here, I repeat, with nothing but workings of the imagination. There are at our disposal no intuitions of infinite divine attributes as Clarke supposed himself to possess in the case of space and time. And how in fact would one have to understand the relation of such divine attributes to the created things that we perceive? Are they supposed to stand in the relation of subject to its properties? Then pain and error and evil would have to be carried over into the Godhead as its properties. But what other relation could it be, since Clarke certainly was thinking of no mere causal relation.

Nothing can help us here other than the renewal and improvement of the entire psychological analysis.

6. Let us examine without prejudice what is given intuitively in sensation, in inner perception and also in memory. It has been said, and is often still said today, that there appears to us here something individuated.[149] Berkeley went so far as to affirm that we are never able to present something merely in general, and he led not a few to

the same nominalistic conviction. In truth, however, just the opposite is the case. As certain as it is that every thinker is individually different, it is equally certain also that no thinker, in being conscious of himself, has his individual peculiarity revealed to him thereby. No contradiction would lie in the idea that two thinkers, in being conscious of themselves as thinkers, would have perceptions which would coincide in every respect. That which individuates lies in the substance. And it is not at all the case, as some suppose, that this is apprehended by us when we perceive ourselves. Rather we apprehend it only in such generality that doubt is even possible as to whether the thinking substance is corporeal or an unextended being.

Things are no different with regard to the sensation of primary objects. Let us take, for example, the objects of the visual sense, particularly favourable to investigation since this sense is distinguished from the others by its special distinctness. Here some believe that we have individual intuitions of determinately localised coloured objects. There may be many red points, but there can be only one red point in the same location, and thus they appear determined individually. On closer investigation however one discovers something very remarkable. Johannes Müller said that the seen objects are 'projected outwards' in the seeing. This is true, but stands very much in need of clarification. If something is projected, then this means here that it is presented as standing at a certain distance. But if this is so, then of course at a distance from something which itself has a place. And what is this in our case? Is it also something that appears qualified as coloured? Evidently not. Petronievics[150] came for this reason upon the idea that it is a point at which we present ourselves, as visual perceivers, as being located.2* It may be more correct to say that we present a certain point without any qualification *in modo recto* and the coloured objects only *in modo obliquo*, i.e. as something from which this point stands apart in a certain direction and to a certain extent. I

2 *Anatomical investigations have incontrovertibly refuted the view of Descartes, which must probably also be ascribed to Leibniz and which Herbart still held on to, according to which a certain individual point of the brain would constitute the seat of the soul, that is, of the subject of our sensation and thinking. The 'life node' ('*noeud vital*') which Flourens wanted to consider as such (with what seemed more reason than in relation to other organs) has shown itself, like the cerebral hemispheres, to come in pairs; moreover, it was possible to preserve psychic life for several days after its total extirpation.

hold this to be the precise description of what is to be understood by 'projection' here.[151]

If we examine what has been said, then, in regard to the question whether or not visual perception reveals to us something individual, everything depends on whether or not this place that is presented entirely without qualifications *in modo recto* does or does not somehow appear as qualified and as quite precisely qualified by an absolute determination of location. And in fact it turns out that it appears to us only quite generally, as a localised point[152] standing to others in certain relations of direction and distance which appear the same for every localised point in relation to every other. However precisely specified these relations may be thought, they can still not yield an individuation. He who knows by how much the fortune of one exceeds that of another may still be ignorant of how much the one or the other owns.

7. Something similar to what holds regarding place is also manifested in regard to temporal determination. We see, as one says, that something is at rest or that something is moving with a certain velocity. It appears to us in the one case as being earlier and later, then and now in the same place, in the other case as being in ever different places. Here the now appears to stand at a distance from the then in the same way as something later from something earlier. But does there appear to us, in addition to this relative determination, some other absolute peculiarity of the now and of the then? Not at all. That which was is now past, and a moment in time which is future will at some time be now, and then it will appear exactly like the present now. Thus exactly the same holds in regard to what is temporal as we have established in regard to space. There is lacking *every absolute differentiation*,[153] though relative determinations are given multifariously. They are given in such a way that there appears one point alone, namely that of the present, as presented *in modo recto*. Every other point appears *in modo obliquo*, which is also why we can say of what was just as little of what is thought that it *is* in the proper sense.[154] For it is characteristic of the way in which things stand temporally at a distance from each other, as opposed to the way they stand apart from each other in space, that only the fundament exists, and not also the terminus of the relation. And thus also the now, while certainly always belonging as boundary to a temporal continuum that begins with it or ends with it or endures through it, belongs always only to a continuum which exists actually entirely and uniquely in this boundary. This has been deemed to be impossible,

because one had been caught by what tends to hold in the case of a spatial continuum. Thus one has supposed that if a boundary were to belong as boundary to a continuum, then this continuum would have to exist as a whole and not merely according to this boundary. Yet that this is not merely possible but in fact actually so, is revealed to us precisely by our experience of the temporal now with immediate evidence. Indeed even a spatial point might exist as belonging to a spatial continuum that existed not as a whole nor in any extended part but only in this its punctual boundary. This would be the case for example if a cone were gradually annihilated from the base up. In the moment of complete annihilation, whether this be final or followed immediately by a new creation in the opposite direction, the uppermost point would be the only one that had not been annihilated, and would indeed belong to the newly arising cone as first point as it belongs as last to the cone that is passing away.

8. Let us say once more: we do not apprehend the specific absolute peculiarity either of the present, or of any other temporal moment from which the present would stand apart as later from earlier or as earlier from later; we apprehend only the relation it bears thereto. Indeed this is all that we are able to bring to presentation here. Just as, in spite of such restricted power of presentation, we can perfectly well say that the present moment reveals itself as at a greater distance from a given moment than from a third, so also we can say that it reveals itself as at the same distance and in the same direction from the temporal points of two events and that it is specifically the same for all that is present. I say the *same* however, not *one and the same*; what exists at the *same* time does not exist in *one* time, just as one can say of two things which are red in the same way that they are of the same colour, not however that they are of *one* colour, and that they are of exactly the same red, not however that they have one and the same redness.

Whoever wanted to assume a time apart from and in addition to temporal things and also a now apart from and in addition to the now existing things would be making a mistake similar to that made by Plato in his doctrine of ideas. Immediate experience does not reveal anything of this sort, and nor either does such a conception suggest itself to us as necessary, nor can it be legitimately inferred. We know only that very many, indeed all things are subject to temporal determinations, even though we are aware of them as absolute only with the utmost indeterminacy and generality (though with manifold relative differentiation).

Something quite similar holds also of spatial determinations. We have none at all that are absolutely specified, but we do none the less possess the presentation of something spatially determinate in general, as also a manifold of determinations of relative spatial differences. Neither the one nor the other is however to be hypostatised, as if there would exist *in addition* to bodies and what is corporeal some unitary reality infinitely extended in length, breadth and depth which, itself immobile, would comprehend corporeal substances and properties within itself and impart to them their absolute local determinations according to whether they are comprehended in this or that part.

It has been said that if a body is to move then there must exist an empty space into which it moves. This is just as compelling as if someone were to say that, if something is to change colour, there must already exist a colour which it then takes on.3* Just as it suffices for a change of colour that something exists that is capable of acquiring colour in such and such a way, so in the case of movement it suffices that something exists that is capable of being determined as to place in such and such a way. Like the body changing colour, so also the body that moves suffers an alteration, though in a different genus of determinations.[155]

Compare also the absurdity that would follow if one wanted to demand the existence of an empty time into which things would enter as long as they persevere — by analogy to the empty space for that which moves. For this would after all have to be an empty future, but in fact nothing that is in the future *is* in the proper sense, and this is why one has to say of the future empty time that it does not yet exist, just as of a future man or any other future thing.

9. Let us now however, after all that we have discovered, look back at the views concerning time and space that were expressed by famous thinkers. We shall certainly not be able to agree with any one of them, but we shall nevertheless find the mistakes that were made understandable. There was no shortage of special temptations in this regard, many a seductive analogy proferred itself, and conflicting deviations from the correct path could be made, so that where someone fell into error he did this by recognising another error as such and striving to keep himself free from it.

When Leibniz declared time to be merely the order of

3* And indeed why not also: if someone is to enter into marriage, then this marriage must already exist beforehand?

succession, space to be merely the order of juxtaposition, this testifies to the correct insight that it will not do to conceive time and space as something outside what has temporal and spatial existence, whether as two things immeasurably vast or, doubly absurd, as two non-things that would serve as preconditions for every thing. Leibniz does not fail to notice, either, that intuition in no case presents to us something with absolute spatial or temporal differentiations but on all sides only relative determinations. This has to be counted in his favour and we therefore want to judge him leniently when he holds that, since experience here offers him nothing but relative specifications, it is only relative specifications that would in truth exist. If with the relative specification of the relation of red to blue there is given to us also the absolute difference of both red and blue, so one might suppose that, were there to exist not merely relative but also absolute specific differences in the case of what is spatial and temporal, then these, too, would no less have to be given in experience. And now because they are not, Leibniz infers that they do not exist at all in reality.

There were also however opponents of Leibniz, such as Newton and Euler, who began remarkably enough by regarding this assumption as not impossible, and it was only on the basis of special experiences that they decided in favour of accepting absolute differences, in particular in the case of space. Euler found their absence in intuition so striking that, while holding as certain the existence of an absolute plurality of places set apart from each other, he claimed for them at the same time an exception to the *principium indiscernibilium*. He did not appreciate the absurdity of assuming as foundation for greater and lesser distances pluralities which are indeed many but not otherwise specified at all. Admittedly the same absurdity is to be found already in Leibniz's doctrine. How can there be distances if it is not the case that all that stands apart is specifically different? The reference by Euler and Newton to special experiences was here not at all necessary for the refutation of mere relativity.

Newton, however, in guarding himself against Leibniz's error, fell into an opposite and hardly less significant error of supposing, as we have seen, that the idea of an absolute space and time existing for themselves is inseparably bound up with the idea of absolute temporal and spatial determinations. More careful consideration shows this inference to be invalid. We have to do rather only with absolute peculiarities of spatial and temporal things, determinations which these can so little do without that the assumption of such absolutes as existing for themselves leads to nothing but a useless duplication. When we refer to a musical tone as c, then we allocate to it a certain position in the scale. But we do not believe that this scale would exist for itself, in addition to this and other similarly sounding tones, as though the individual sounding tones would correspond with elements of the scale and be conditioned thereby. For then the tone would be given twice, once as element of the scale and again as an individually sounding tone. But just as little are we allowed to fall into the error of believing that, for a body to have a spatial peculiarity, and for any arbitrary thing to have a temporal peculiarity, it must be the case that the same peculiarity applies also to some element of another thing or non-thing, which would enduringly and necessarily comprehend this element together with infinitely many others — and in such a way that this alone would make possible participation in the given peculiarity, precondition it in some mysterious fashion. Exactly the same peculiarity which would constitute the element of this fictitious thing or non-thing is also supposed to be enjoyed by that which it is said to condition. It is distinguished therefrom through nothing other than negative determinations and through the assertion that it is immediately necessary and eternal, as Aristotle said of the 'ideas' of Plato and of their relation to the corresponding individual things. Indeed, a sphere in itself would not be more spherical than a genuine individual sphere, and red in itself would be no redder than an individual that were genuinely red.

10. It is perhaps not superfluous to ward off certain further objections to what has been said here concerning space and time. The view that when we have to do with what is spatial and temporal we have to do with nothing absolute but only with what is relative finds wide acceptance today among the natural scientists. Many say that one is not allowed to assert that something is if it is to be found nowhere in experience. Now we ourselves have admitted that there is given to us nothing but merely relative spatial and temporal determinations. Thus one would have no right to conceive anything as vested with absolute spatial or temporal peculiarities. But someone could argue exactly similarly that no intuition shows us something individuated, and thus one would not be allowed to assume that individual differences exist at all, but must regard as real only universals as universals. As far as I know, however, there is no one, excepting perhaps in a certain sense old Parmenides, who has fallen into this error, and even Parmenides no

longer conceived being as something general but rather as something individual because single. Otherwise it was held only that there are both individual things and also universals, and this error, too, has long ago been refuted. If I have learned that someone possesses a dog, then I certainly cannot thereupon infer that he possesses a mastiff, or a pomeranian, or a poodle, or a dog of any other sort precisely known to me. I can safely infer only that it is not merely a dog in general but a somehow specific sort of dog. And the same is to apply to every other universal.

We have to judge likewise, now, in regard to the existence of any absolute determinations where only relative determinations are given. If I know that someone has twice as large a fortune as another, then I may be ignorant of the absolute size of his fortune, but that it has some determinate size is clear to me none the less. Whoever would deny this would fall straight into the error of those ultrarealists who teach universals as existing for themselves. For the fortune that is twice as large can be owned just as well by someone who possesses millions, thousands or only hundreds. The most various of absolute specifications are conceivable, but the absence of every specific absolute magnitude quite inconceivable. Thus the assertions on the part of the physicists of the mere relativity of what is spatial and temporal have always appeared to me to be grotesque.

Still more peculiar, however, is the fact that Leibniz, the philosopher who had rejected non-individuated universals as unacceptable already in his doctoral dissertation, fell into such an error and noticed none of the obvious absurd consequences thereof. Thus according to him God could just as well have created a world without as with a beginning, yet as between a later or an earlier beginning he had no choice. If, however, one had asked Leibniz whether it is in God's power to end the world and to have another begin one thousand years later, then he would have had to answer this question in the affirmative. For then there would be given a relative temporal determination in the temporal distance between the two worlds. Absolutely, however, the progress of the second world would perhaps be distinguished not at all from that of the first, so that, with the simple falling away of the first, the second would appear as having been transformed into it. Its beginning point would have become the beginning point of the first, its end point the end point of the first, which however had already been 1000 years earlier and which is no longer possible at all. Leibniz, the great enemy of all indeterminacy, falls here unwittingly into the absurdities of indeterminism. And Clarke in his correspondence correctly saw himself as having been in this respect the victor.

11. Meanwhile however one must warn against another misunderstanding. If we say that in regard to time and space we have to do not merely with relative but also with absolute peculiarities, then we do not imply that a temporal or spatial point could exist without any connection to any others.[156] Thus we do not believe, either, that someone who thinks something, because he himself exists absolutely as thinker, would lack the relative connection to something as thought. Spatial and temporal points which exist in reality could never be without any spatial and temporal relation. Such would amount to a blatant contradiction. They exist after all always only as boundaries, and therefore as belonging to certain continua and as standing apart to a greater or lesser degree from innumerable other spatial and temporal points, whether these now exist or not. What we maintain, therefore, is not the absence of relative determinations. Rather we protest only against the doctrine of absolute indeterminacy, which deals with spatial and temporal points in the same way as someone would deal with the whole numbers who asserted that there could exist such a number which has only specific relations of magnitude to other numbers. Such a number would possess no absolute specific numerical determinations of its own, being neither 1 nor 3 nor 100 nor any other specific plurality, but a quantity only relative to others, equally specifically undetermined, so that all of them would be, absolutely speaking, only numbers in general.

12. Modern physicists have also come up with the following strange idea. After denying the existence of all specifically different absolute spatial and temporal peculiarities and allowing only relative specifications, they go on to confound this relation with that in which one who is sensing stands to what he senses, or with that relation in which someone who has a phenomenon stands to this phenomenon. Not merely should it be the case that no absolute temporal and spatial peculiarities should exist, but no relative specifications should exist *in themselves* either. These *relativa* should no longer have existence for the things in themselves, but only for the things as they appear to a perceiving subject. Thus the same thing, according to who is perceiving it, could appear to be at one and the same time spatially and temporally nearer or more remote, more or less extended — all of which is rendered impossible by what was said earlier against Kant.

13. And yet another strange doctrine has arisen according to

173

which time would be nothing other than a fourth dimension of space. Being spatial and being temporal are supposed no longer to be different concepts at all. A unified concept — we might call it that of the spatio-temporal — of something having four dimensions is supposed to take their place. Much has been made of this supposed discovery and its proponents do not feel themselves in the least animated by the question why, then, the distinction between time and space was ever made, and was made by all the peoples known to history. The reason, in truth, lies in the fact that the difference is obvious to everyone, and is seen already in the fact that, in regard to time, nothing that *is* can stand apart from anything else that *is*. Only what is in the present is in the proper sense to be accepted or acknowledged, and nothing in the present stands apart temporally from anything else in the present, but only from something past or future, of which the one no longer is, the other is not yet. Therefore the last two are not to be accepted in the proper sense but can be acknowledged only *in modo obliquo*, in which manner also something can be acknowledged as thought, even though what is thought (for example a round square) is perhaps absurd and in the strict sense impossible. Things stand quite differently in regard to spatial distances: of two towns lying at a distance of 1000 miles apart, the one is to be accepted in the proper sense just as much as the other.

That this confusion of the spatial and the temporal constitutes no glorious advance but rather a deplorable step backwards in relation to what had been generally accepted, is sufficiently clear already from the fact that something could perfectly well exist without spatial extension and position and without in any way belonging to something spatially extended, namely as a mental substance, or also as belonging to a topoid of any arbitrary number of dimensions, while nothing could exist except as belonging to something temporally extended that proceeds in one dimension.

In short, the relation of before and after may certainly bear a number of analogies to that of spatial distance, it is however undeniably not identical therewith. If we set up three coordinates perpendicular to each other in space, then it is quite immaterial in which direction we would have them individually proceed. And if we want to have time truly count as a fourth spatio-temporal coordinate, then it too must admit of being displaced without difficulty — and indeed allow itself to be linked just as consistently and necessarily with the displacements of the other axes as do they amongst themselves. The spatial point that is here is not identical with the temporal point that is now, and a spatial there is not identical with a temporal then, even if the same thing is here and now, and there and then. The spatial point is a boundary for something spatial in many conceivable spatial directions, perhaps even in all, but it is never a boundary in either of the two temporal directions, and the converse holds of the temporal point.

Just as a perpendicular can be erected from a point in a plane, so, if time were the fourth dimension of space, could such a perpendicular be thought of as being erected in a present three-dimensional flat space, which would then be a time of a certain length. This time would form a right angle with every one of the straight lines proceeding from the base of the perpendicular, and one could imagine a hypotenuse having been drawn from an end point of one of these lines to the tip of the temporal perpendicular which would then perhaps have to be conceived as similar to a motion which varies continously as it were both as to location and as to time. These would not, however, be two variations, but one variation of the same degree, a degree which would apply to the purely temporal line and to the purely non-temporal line in space whose end points it joins. It is so difficult to think oneself into all of this because it is already absurd to have constructed a non-temporal spatial line as forming a vertex with a non-spatial temporal line as otherwise spatial line comes into contact with spatial line and forms thereby a broken longitudinal magnitude.

None the less, however, the fiction that time is the fourth dimension of space will reveal itself as harmless in a number of respects. This was stressed already by Lagrange, who found it admissible that mechanics should treat time as a fourth dimension of space. It can indeed be shown that there must obtain a precise relation of magnitude between intervals of time and the lengths of lines and that for example it can be said without absurdity that the present is at precisely one meter's distance from a certain point in the past (though one must not imagine oneself able, by measuring back into history, to find out which one).

14. What, now, after all that has been positively established and polemically rejected, is to be said of what is temporal and of what is spatial?

Above all that we have to do in both cases with something real.

And moreover with something that is always absolutely determined, specified and individuated in reality, even though it is so determined for us in relation to none of its absolute specifying

175

marks. It is something that is capable of being thought intuitively and conceptually only with relative specifications. What is temporal, like what is spatial, is available to us only as to its *most general* absolute concept.

As regards this concept we can say that both what is spatial and what is temporal appear as something that can exist only in the context of a continuum. Their ultimate elements are boundaries which are as such nothing for themselves but exist only as belonging to a continuum. More specifically, the spatial boundary belongs always to a three-dimensional continuum, the temporal always to a continuum of one dimension.

This continuum coincides in the case of what is *spatial* with the corporeal as such. To this there belong both what is accidental and what is substantial. If we see something coloured, then as already said we project it, in other words we present to ourselves a point as standing apart from it in a certain direction and distance, without however presenting this point itself as coloured. We may well say that we present this point as standing to a coloured thing thought to be just there in the same way as to an accident of a subject that is homogeneous to the fundament of the relation in question.[157] And thus we are allowed to say that something spatial is as such not merely something real, but also something substantial whose substantial determination then also enters into the determination of the accidents for which the substance is the subject.

Compare, now, what is *temporal*. Just as what is spatial coincides with what is corporeal, so what is temporal coincides with things as such. And just as it is impossible to present anything other than what is real, so also it is impossible to present and therefore also impossible to accept anything other than what is temporal. Hence it is established that nothing is other than what is temporal and more precisely what is present.

15. A number of objections will certainly be raised against this, but they can be easily refuted.

a) Thus one could say that even of what is temporal itself there is much that is accepted not as present but as past or future, such as historical facts and astronomical and many other events that we can foresee.

Precisely considered, however, the acceptance of something as past is just as little acceptance in the proper sense as is the acceptance of something as thought. Just as in the latter case what one properly accepts is a thinker, so in the former case what one accepts in the proper sense are actually existing things as standing

176

apart as later and earlier — and then, in accordance with the peculiarity of standing apart in time, that from which something stands apart *is not* in the same way in which this something itself *is*.

b) Another objection may point out that there are laws of such a generality that they exclude all temporal qualification. Such laws are supposed to have eternal, not temporal, validity. Thus for example the 'is' in Pythagoras' Theorem could not be interpreted as an 'is presently'.

It is easily seen, however, that when dealing with the relationships of squares one has to do with relations that belong to temporal things. And this may be seen still more clearly when one considers that just as one speaks of eternal geometrical laws so also one speaks of eternal laws of mechanics, as for example when one affirms the law of inertia to be one such. It is however obvious that these laws, in relating to motion and rest, relate to what can exist in no other way than temporally. If this is acknowledged however, then looking back at the answer to our first objection will supply the answer here also. Pythagoras' Theorem says nothing other than that what is in the present can neither comprehend a right-angled triangle lacking in the equality of the square on the hypotenuse with the sum of the squares on the other two sides, nor stand apart from such a triangle in which this equality has been or will be lacking.

c) There is yet a third objection that cannot be left out. Eminent thinkers in ancient as well as in modern times have expressed themselves to the effect that, if not several, still there must be at least *one* being which, as immediately necessary, is removed from all change and is thus also without any sort of temporal before and after. Aristotle affirmed this of the divinity and indeed he ascribed timeless existence also to the spirits of the spheres and to the substance of the spheres themselves, which he held to be unchangeable. The Neoplatonists, too, taught the timelessness of their primeval One, and the speculative Fathers of the Church, above all Augustine, likewise declared God to be absolutely without change and therefore lacking in all before and after. This doctrine then maintained itself in the middle ages and we find it again even in the work of Leibniz. The English thinkers Newton, Clarke and Reid who subscribed to Christianity certainly acknowledged the eternity of God as a before and after without beginning and end — yet they, too, came close to the idea of God's timelessness in a peculiar and in my opinion absurd fashion, since according to them every instant of the duration of God is supposed to be *without difference* from every other. If, with Leibniz, one demonstrates the contradiction that lies in this, then this may lead once again to the denial of God's temporality. Of those who do this, however, many have at the same time disputed that God would fall with other things under the concept of what is real. To God there would apply no concept in common with what we are able to present to ourselves, and thus even after the exclusion of the divinity from what is temporal it would still remain thinkable that one should identify the two concepts of what is real and what is temporal.

Such far-reaching agnosticism has however exposed the whole of theism to the ridicule of Hume, who remarked — and not without plausibility — that the theist who adheres to it is no longer distinguishable at all from the atheist. And thus also we see Leibniz very eager to affirm that God is truly something real, and palpably also that he is something knowing and loving and preferring what is good. Since he still, in spite of this, denies all temporal progression in God, we therefore find him denying once more the identity affirmed by us of what is real with what is temporal. Clearly he had not made clear to himself the consequences of this. Trendelenburg, a great admirer of Leibniz, declared it as reasonable that, although perhaps rest could result from motion, motion could never result from rest, a proposition that becomes still more certain if we interpret it as meaning that something completely unchanging cannot serve as cause of something that changes. Now however there is change in the world. Hence also the first principle of the world must have a changing existence. Aristotle, who taught the opposite of this, nevertheless saw better than Leibniz the difficulty of such a doctrine and took great pains to bridge the distance between what is entirely changeless and what is subject to multifarious change. The first change is supposed to consist in a rotation which, proceeding with constant uniformity, always makes good what it has abolished. It can easily be shown that this artificial Aristotelian construction is inadequate, for his uppermost sphere of the fixed stars as surely does not put like in place of like as it does not always give light and as the fixed stars shining from it are distributed irregularly. But his explanation is inadequate also in other respects. His first moving principle, because it cannot be a motion, is supposed to be an all-knowing and ordering intellect. But this intellect — even if the entire order of the world were known to him in its progression — would not in truth be all-knowing should there remain hidden to it what in this world order is, or already was, or is yet to be. In other words, as certain as it is that the world changes, it is certain also that truth relating to it changes, too, and since the omniscient being knows all truth, but can know nothing over and above this, it follows that, in so far as it knows, it must be caught up in a change just as certainly as is truth itself, i.e. as the things are constantly changing. Far from entering into contradiction with itself, however, the omniscient being successively holds something to be future, present and past, and the one whose omnipotent will determines everything successively wills something as future and as present and then goes on to find good what has earlier happened. It is therefore apparent that if we identify what is temporal with what is real then we must not thereby permit ourselves to be forced into any vain agnosticism. He who includes God among what is temporal because he includes him under what is real, is guilty of no inconsistency, since it is much rather the case that, just as God cannot be without being something real, so also he cannot be the principle of a world subject to alteration without his life consisting in a temporal progression. And this, far from doing harm to his immediately necessary perfection, is much rather precisely what makes this perfection possible and allows it to appear in harmony with itself.

Thus also the last objection fails. The thesis that has hereby been secured against doubt does however provide the explanation why, in spite of the extravagant judgments of many philosophers, laymen are agreed in their healthy common sense that being and being present amount to one and the same.

16. Do we not however come into contradiction with ourselves when we grant of all that is real as such that it belongs to what is temporal yet at the same time deny time as a fourth dimension of what is spatial — for after all would not everything spatial be something real and thus also something temporal?

This has still to be considered. We shall probably have to look for the answer in the fact that the concept of what is temporal is superordinate to the concept of what is real and thus comprehends within itself species lacking in all spatial continuity. (Even the concept of a sensory quality, which includes as species the concept of colour, has in itself for this reason nothing of the character of colour-differences as such.)

The complete elucidation of this point is however to be achieved only after entering into yet other important ontological questions,[158] questions with which already Aristotle had concerned himself, though without, in my opinion, having answered them successfully in every respect. He declared that there could never, in defining, occur an overlapping of differences, that much rather every definition, from the highest generic concept down as far as the last specific difference, proceeds monostoichetically (universally in one series). The last specific difference is equal in its content to the whole of the definition. On this uniformity of the series there should rest, then, the uniformity of being.[159] But this doctrine is untenable. Just as Aristotle was in error when he would not have a plurality of reals count as being itself a real thing, so also when he held that all plurality of equally immediate differences would be excluded from the concept of something unitarily real. And just as there is possible in the case of other concepts a plurality of not merely superordinate but also of equally immediate specific differences which belong to different series of specifications, so also is this possible already in the very case of the all-inclusive concept of what is real.[160] We said that this coincides with the concept of what is temporal and that everything real which is, is in the present. We said also that it appears as differentiated from what was or will be in no absolute specific determinations, certainly however by relative differences, intervals, which point to absolute specific differences that are completely transcendent to our presentation. In these latter, therefore, we have to acknowledge the specific differences which apply immediately to what is real in a series by which it becomes specified. It is in relation to this series that we describe the real more particularly as something temporal. The real is something that specifies itself temporally. This does not however exclude its manifesting immediately specifying differences also in other series. Such a difference may be found in the fact that some real things are *collectives* of real things or are plural real things, others however are isolated, are *singular* real things.[161] And another such difference may be found in the fact that some reals are *substantial*, others however *accidental* and such as to belong to this or that accidental category.[162] Moreover one may want to seek out such an immediate differentiation in the fact that when we have to do with what is real we have to do sometimes with what is *spiritual*, sometimes with what is *corporeal*, with what belongs to something spatially extended, and on yet other occasions with what belongs to a *topoid* of another number of dimensions. And indeed the question could be considered also whether one should not declare as immediately differentiating for what is real in general certain differences according to which what is real is real either *for itself* or only *as belonging* to some other real that is. For even if we were to

180

find that nothing real could be, other than as boundary and as belonging to something real — because it would otherwise not be temporal — this would still not be a matter of belonging to something else that *is* except in this single boundary itself. Still, perhaps these many-sided specifications do not have an equal right to count as immediate specifications of the concept of what is real in general, and we do not here want to get involved in an investigation of what is to be said in this respect about the differentiation between the spiritual, the corporeal, and topoids of other numbers of dimensions. And, whatever may be decided in this respect, one thing has, I think, been clearly shown: the fact that everything corporeal, everything belonging to something three-dimensional, belongs also to something temporally extended, should not be allowed to count in favour of the idea that time is to be seen as a fourth dimension of space. Thus there falls away this doubt against what was already positively established above.[163]

Notes

by Alfred Kastil

1. The issue as to whether concepts are innate in us Brentano held to have been long ago decided negatively. Aristotle, Locke, Hume and he himself have in fact said all that is necessary in this regard. Cf. F. Brentano, *Versuch über die Erkenntnis*, ed. by A. Kastil (1st ed. Leipzig, 1925; repr. Hamburg, 1970), Part I, pp. 26ff., 40f., 153f.

2. Cf. F. Brentano, *Psychologie von empirischen Standpunkt*, (2 vols., Leipzig, 1924/25; repr. Hamburg 1971), pp. LXXXV-XCIII, Eng. trans. as *Psychology from an Empirical Standpoint* (London, 1973), pp.405-408 (introduction by the editor on inner perception and observation).

3. Cf. *Psychologie*, vol. II, pp. 145ff., Eng. trans. pp. 281ff. Along with the thesis tht all our concepts are either abstracted immediately from intuitions or obtained by combination out of marks so obtained, Brentano also rejects all 'a priori conceptual categories of reason'. In this sense we are happy to allow him to count among the 'psychologistic thinkers'. Cf. *Psychologie II*, p. 179, trans. p.306.

4. H. Poincaré, *La Science et l'Hypothese* (Première Partie, Chap. II). Cf. also F. Brentano, *Versuch über die Erkenntnis* (2nd enlarged ed. 1970), pp. 209-212, 233-236.

The Addendum on pp. 39-44. consists in a critical consideration of a construction put forward by Dedekind. Dedekind was perhaps however concerned not so much with a derivation of the concept of continuum as with the provision of a sign-system so constituted that through it every attainable point of a continuum could be named.

Regarding Poincaré, he seems to have failed to see that all our sense qualities include already from the start the datum of place. The idea that the concept of the continuum was non-intuitive would however be attractive to an extreme empiricist who wanted to have space derived only from quality. In an essay 'Why space is three dimensional' (in H. Poincaré, *Dernière Pensées* (Paris, 1913), p. 77, Eng. trans. as *Mathematics and Science: Last Essays* (New York, 1963), p.34, we read: 'It therefore seems that we cannot construct space by considering sets of simultaneous sensations, and that we must consider series of sensations.' And further on pp. 79f., trans. p. 36: All that is required for the construction of a physical continuum is 'series of muscular sensations'.

5. The text should say more precisely that the same place could not be presented at the same time with two qualities as in the present, *assuming that it is presented with both in recto*. This qualification is then added in Brentano's note. For a complete understanding one should consider also the following: Already he who thinks something as thought thinks with a *modus obliquus*; he who thinks something as thought in the past, however, thinks it with a second *modus obliquus*, for the *modus praeteritus*, too, is a *modus obliquus*.

6. Cf. F. Brentano, *Untersuchungen zur Sinnespsychologie* (1st ed., Leipzig, 1907), pp. 51ff., (2nd ed., Hamburg, 1979), p. 66ff.

7. Brentano is thinking here of the temporal dimension that applies to the body in so far as it moves or continues to exist. But this temporal extension applies to that which is zero-dimensional also, and is not to be referred to as a fourth dimension of what is spatial. Cf. Section 17.

8. This concept of plerosis (which is related to the mathematical concept of neighbourhood [*Ufer*]) is, along with that of teleosis, one of the most important in Brentano's synechology (theory of the continuum). The Greek πλήρωσις signifies fullness; a greater or lesser plerosis therefore signifies more or fewer directions in which a boundary is related to its continuum. In earlier works Brentano had spoken of holoclery and meroclery (fullness and partiality of participation).

9. Hence in the diameter of this doubly bisected sphere there coincide four lines having 1/4 plerosis. (Coincidence through contact as contrasted with coincidence through continuity.)

10. To what is said in the text concerning plerosis we add here as a note a dictation entitled 'Plerosis' [*Meg 15*]: 'There are certain cases which seem to present difficulties for the doctrine of the plerosis of a boundary, and in particular also of a point. Thus for example the case where, of a number of lines running in parallel to each other, one is turned in the plane. Conceived as infinite it then cuts in its direction the others from the first moment, but in no first point. If I imagine the initially resting parallel as being moved from a certain moment, then the motion seems to have only a *terminus extra* as beginning, and this is the line at rest, where according to the doctrine of plerosis it ought also to have a *terminus intra* which would coincide with the *terminus extra*.

Yet this *terminus intra* is also truly present, albeit not as a line that would cut the remaining parallels but rather as a line which is co-determined in its peculiarity by the subsequent continuity of succeeding intersections and is thus essentialy different from a line which would constitute the beginning of a plane at rest. If there should be no preceding rest at all, but a turning motion were given already from the first moment of creation, then the *terminus extrinsecus* would be entirely lacking, and thus it must be undeniable to everyone that a *terminus intrinsecus* cannot be lacking here. Likewise also it is the case that this, although not intersecting the parallels, is yet characterised as the beginning of the divisions following continuously upon it and characterised differently according to whether these vary more slowly or more quickly.

Even the case where opposite motions follow upon each other serves to lend support. It is clear that Aristotle and the scholastics were incorrect when they declared this to be quite simply impossible. It is given wherever a body thrown up in a straight line then falls straight down, though with infinitesimal decrease and increase of velocity. A moment filled with rest cannot however be assumed, but rather only one such that the beginning of the sinking coincides with the end of the rising. The body moves in this moment taken for itself alone neither upwards nor downwards. Taken in conjunction with the preceding and following moments however it both rises in it and falls.'

11. One should not suppose that this interpolation was provoked by A. Einstein's example of the discs in *Die Grundlage der allgemeinen Relativitätstheorie* (Leipzig, 1916, § 3), for Brentano did not become acquainted with this part of the theory of relativity. The sense of the passage is this: an axis is a boundary and as such nothing for itself; hence it is obvious that what is bounded, if it rotates, rotates also in its axis, i.e. that even this continuously alters its direction, for it is in this, after all, that absolute rotation consists. If one imagines the sphere as made up of four parts — of gold, silver, copper and tin — and as turning, then there coincide in the axis four diameters in contact with each other, each having one quarter plerosis, and one can say that the axis in that quarter plerosis that is gold, is now pointing in the direction in which formerly the silver quarter was directed, and so on. Analogously, also, if for example one of the quarters is missing.

12. A measure of time presupposes after all a measuring soul.

13. Cf. *Psychologie II*, pp. 200ff., Eng. trans. pp. 311ff.

14. Brentano calls these four-dimensional formations topoids and he occasionally refers to the three-dimensional surface of a four-dimensional topoid as an '*Oberkörper*' (superbody).

15. Here A. Marty is intended.

16. One could here assume also a thinking *in obliquo* and say that the wish that something should happen or be going to happen would include, if not the judgment itself, then at least the presentation of a judgment in which something is accepted as present or as future.

17. Just as Brentano says of two judgments, one of which accepts what the other rejects, that they have in regard to the same object a different content, so also he says of two presentations, the one presenting as present what the other presents as past, that they differ in their content.

18. This remark is directed polemically against Anton Marty, *Raum und Zeit*, ed. by J. Eisenmeier, A. Kastil and O. Kraus (Halle a. S., 1916).

19. Cf. F. Brentano, *Vom sinnlichen und noetischen Bewusstsein*, ed. by O. Kraus (1st ed., Leipzig, 1928), pp. 84, 113, Eng. trans. *Sensory and Noetic Consciousness* (London, 1981), pp. 61f., 82f. (The 2nd edition of this work appeared in 1968 under the title *Psychologie vom empirischen Standpunkt. Dritter Band. Vom sinnlichen und noetischen Bewusstsein*, and will therefore be cited in what follows as *Psychologie III*.)

20. Thus it is clearly asserted that for Brentano there are as many temporal continua as there are things. There is no place in his ontology for 'time' as unitary entity (or non-entity). It is for this reason that — in contrast to what has been asserted by C. Stumpf and E. Seiterich — there can be drawn no pantheistic consequences from his doctrine that even the being of God is a temporal progression. There is no thinker who stood further from any concession to pantheism than did the theist Brentano.

21. Brentano speaks of continuous transitions of colour only as a fiction that may be of assistance in gaining intuitions. The intermediate colours are according to him in fact colour mixtures. He goes on below to distance himself expressly from the usual conception.

22. For a better understanding of what is at issue in this paragraph and in the preceding § 15, it may be of use to cite a dictation from 21 November 1914 *[Meg 22]*, which explains the distinction between primary and secondary continua as follows: 'I say that a line would be in regard to its length a primary, and in regard to its direction a secondary continuum. What comes into consideration for the length is only the quantity of transitions from place to place. If in two lines I distinguish to infinity ever smaller and smaller distances always equal in size in the one as in the other, then that line is the longer in which the number of such distances is ultimately the larger. The variation from place to place has here no greater or lesser degree, and immediate neighbours are always specifically different as to their place. One can say the length of the line is equal to the size of the manifold of differences of place. Things stand quite differently in the case of the direction of the line. It has in every point a direction and thus one can say that it has a direction continuously. It is possible that the direction it has in one point is specifically completely the same as the direction it has in another. But it is no less possible that it should change direction in some single point, indeed that its direction should vary throughout, as happens in the case of lines that are curved. And where it varies, it can vary more or

less intensely, sometimes with a sudden variation of finite magnitude, sometimes infinitesimally — and this, too, in very different degrees.

Another example is something that exists throughout a given time. Here there is to be distinguished the length of the time in which it exists and in virtue of which it appears as a primary continuum subject to a completely uniform change. However it can be the point of a body which is either at rest in the given time or moving more or less quickly and with greater or lesser regularity. In regard to the change of place to which it is subject during this time it is also a continuum, but a secondary one, and here again one can distinguish a continuum in respect both of the variation in distance covered and of the variation in direction. But the body in question, too, during this time, could either remain qualitatively the same or in various ways — indeed continuously — change, and this now with changing degree and direction of qualitative alteration, now without such variation. All of this would therefore be a matter of secondary continua.

Should we imagine for ourselves also a surface that were either uniformly red or varying uniformly or non-uniformly and more or less rapidly from red to blue, then we should have in the surface itself as spatial magnitude a primary continuum, and there would be to be distinguished in it one secondary continuum in so far as it is straight or broken and another in so far as it is coloured.

If we look more closely at the examples of secondary continua that have just been mentioned, then we find that certain of them can arise only in the case of boundaries which are themselves not complete continua. In the case of a line and a surface it can be a matter of change of direction, whilst in the case of a body this is completely excluded since it much rather participates, part for part, in all directions that are in any way possible for bodies. However, in regard to the quality of a body and in regard to the rest and motion of the same, it is clear that what we have to do with here cannot count as mere boundaries.

(Strictly speaking one would, I believe, have to say that even the primary continuum of spatial extension appears only as existing in a boundary; for if I consider it not just as momentarily given but as it continues to exist in time, then either it endures unchanged in case it is at rest, or it varies, in case it is in motion, and thus it appears as a secondary temporal continuum.)'

23. E.g. an act of conceptual thinking.

24. In the case of the red radius the red in the point of the circumference is with a teleosis precisely twice as large as that in the mid-point. For the variation of red through violet to blue would after all be slower on the circumference as towards this mid-point; hence the radius, as boundary of what is variable, participates in this difference of variation. In applying this fictitious example to the real example of motion we read in the dictation 'On the Possibility of a Continuum' of 1914 *[Meg 25]*: 'In the case of lesser teleosis the totality of points of the larger course coincides with the same quantum of temporal points with which in the case of larger teleosis there coincides the totality of the points of the smaller.'

25. The passage set by me in parentheses should not be allowed to mislead. One could speak of full teleosis only to the extent that one disregards the context of the red boundary line of the rectangle. That is however abstraction in the sense of a fiction. Thus also we can 'abstract' from the temporality of what is spatial and this contradiction may, because it leads to a simplification, prove of service for our purposes.

26. This thesis would be, in the jargon of the modern Poincarists, 'senseless' (*un mot vide de sense*), according to the dictum: What cannot be measured, cannot be regarded as existing.

27. As one can infer from what follows, other determinations are required in addition to this 'at the same time'. To the simultaneity one must add also the equal-sidedness of the plerosis, and differences in teleosis must be taken into account also. Otherwise something could be and not be at the same time. Thus there is to be corrected the error of Kant, who, in order to maintain the law of non-contradiction as an analytic judgment, found it necessary to exclude from it the concept of time (where time as form of inner sense is involved we have, according to him, to do with synthetic judgments). Account must also be taken, however, of a number of differences to which even the otherwise so careful Aristotle had paid no heed.

28. If one understands by point one that is in partial plerosis, then it is to be accepted, if one understands one in full plerosis, then rejected.

29. Cf. F. Brentano, *Vom Dasein Gottes*, ed. by A. Kastil (Leipzig, 1929; repr. Hamburg, 1968), pp. 423ff.

30. Evidence for this dating: fine handwriting, and thus a longer time before Brentano's going blind. Moreover, the part dealing with modern philosophers ought, on the basis of a remark in a letter to Marty of 12 February 1903, to be ascribed to this time.

On the history of the problem of time see also *Psychologie II*, pp. 262ff., Eng. trans. pp. 358ff.

31. Namely that time is something external in relation to the things.

32. The question is, after all, how something simply lacking in all change can differ from some like thing that is equally changeless in the fact that it *is* at a different time.

33. Intended here is the continually recurring process of being effected by God.

34. The idea is developed in the following criticism of Locke.

35. Just as two bodies of the same volume can differ in their place, so two things of the same duration can differ in their time.

36. For he has demonstrated no continuum of specific temporal differences.

37. Add: 'even in the case of changeless perseverance'.

38. Add: 'not those of their absolute temporal positions'.

39. Cf. *Untersuchungen zur Sinnespsychologie*, 1st ed., pp. 71, 160, 2nd ed., pp. 82, 91.

40. If the duration of a rest is equal in length to that of a simultaneous change of colour or place, then this length must be constituted out of differences of a different sort from those of colour and place.

41. Cf. the criticism of Hume's concept of cause in F. Brentano, *Versuch über die Erkenntnis*, Section 1, Part IV; *Vom Dasein Gottes*, pp. 129f.

42. Cf. Brentano's criticism of Kant in *Versuch über die Erkenntnis*, Section 1, Part I; *Vom Dasein Gottes*, pp. 77ff.

43. Namely in relation to these colours.

44. That is, now more quickly, now more slowly.

45. Which it is, according to Brentano, for perseverance itself is for him a continuous change.

46. Intended is: no positive concept of what is infinite.

47. Because — according to Kant — the whole of time is supposed to exist together in all its parts, as in the case of space. What is past and what is future are supposed to be exactly as what is present, though in other parts of time.

48. For as principles of knowledge there come into consideration only judgments which are evident, which — as Kant himself concedes — his synthetic judgments [*Erkenntnisse*] a priori are not. Cf. *Versuch über die Erkenntnis*, Section 1, Part I (Nachtrag); *Vom Dasein Gottes*, pp. 79ff.

Notes

49. I.e. to the temporal substance there would apply a here and a there, to the spatial substance a before and an after.

50. Namely of the changeless persistence of the infinite line of time.

51. Since after all it should, as a whole, exist in all its parts.

52. The *differentia individualis* are distinguished from the *differentia specifica* in the fact that each can be exemplified only in one thing. Each thing must have its own, likewise unique, individual difference. The genus to which these individuating marks are related as last differences can therefore have no fewer such last differences than there are individuals which belong — or can belong — to it.

53. Which holds for Brentano, but cannot be admitted by Schopenhauer because he takes time to be a unity.

54. Namely for the created things.

55. If, that is to say, temporal connection is supposed to signify nothing other than conditioning and being conditioned.

56. B. Bolzano, *Paradoxien des Unendlichen* (Leipzig, 1851; Philosophische Bibliothek edition, Leipzig, 1920; revised, Hamburg, 1975), Eng. trans. *The Paradoxes of the Infinite* (London, 1950).

57. Cf. F. Brentano, *Kategorienlehre*, ed. by A. Kastil (Leipzig, 1933), p. 26, Eng. trans. *The Theory of Categories* (The Hague/Boston/London, 1980), p. 29.

58. Simply in that it itself continues to exist. One recalls the related question whether, with the destruction of everything present, everything past would also cease to have existed. Cf. A. Kastil, 'Zeitanschauung und Zeitbegriffe', in *Naturwissenschaft und Metaphysik* (*Veröffentlichungen der Brentano-Gesellschaft*, Brünn and Leipzig, 1938), p. 101.

59. I.e. it would have to be the case that, if no heat death is to occur, then every time the quantity of heat had reached a maximum there would immediately follow a reversal of the process.

60. In order to avoid the consequence that the quantity of heat would then at the same time have decreased and not decreased.

61. This is later corrected: see the next chapter.

62. The sense in which the *modus* of the present can be undetermined will be specified under VII. below.

63. This remark serves here as an example of the way in which the *modus* of the present in inner perception can be two-sided, i.e. can be a boundary in both temporal directions.

64. What is intended here is its exact relative determination, for full determinateness would imply that the relevant absolute temporal difference would be intuitive to us. It is however completely transcendent as far as we are concerned.

65. Cf. Addendum under 3 (pp. 77 above).

66. Cf. *Psychologie III*, pp. 84f., Eng. trans. pp. 61f.

67. The idea is this: If exactly similar things are at different positions in time, then these positions in time themselves must be somehow distinguished from each other. And naturally not then in their different positions in some further time which would comprehend time within itself. Temporal perseverance must therefore already in itself amount to a change.

68. It would however not be acceptable to designate time as fourth dimension of space — and already for this reason: that what is non-spatial, too, can be temporally determined.

69. Since however Brentano teaches that we have acquired the concept of what is temporal from a continuum of *modi* of presentation (or also of acceptance) given simultaneously in its various parts, there arises the paradox that the concept of what

187

is temporally continuous has to be taken from the intuition of what is spatially continuous. This consequence would have to lead to a further development of the theory. The beginnings thereof are to be found where Brentano identifies the general concept of thing with that of what is temporal. (Cf. *Psychologie III*, p. 120, Eng. trans., p. 87.)

70. Cf. Brentano, *Kategorienlehre*, p. 68, Eng. trans., p. 58.

71. No indication is at this stage given of its subordination to the *modi obliqui*.

72. Here it is assumed that even what is accepted as past is accepted with a *modus rectus* — only not with the *modus praesens*. As Brentano later held, however, only what is now in the present is accepted *modo recto*; what is past is accepted *in obliquo*, so that 'there was something 100 years ago' would *mean* nothing other than 'there is something which was preceded by something 100 years ago'. We have here not two propositions, of which the one would have to be inferred from the other, but only two different ways of expressing the acceptance of what is in the present as something lying 100 years apart in the sense of the later from the earlier.

73. This paragraph is taken from a dictation of 24 January 1915 *[T 19]*.

74. In a dictation of 24 January 1915 *[T 19]*, where the peculiarity of succession as contrasted with juxtaposition in space is likewise emphasised, we read: 'The order of right and left and of left and right are equal in value, those of before and after and of after and before however by no means. In the case of before and after it is only the one that can exercise an effect through contact, the other not at all; in the case of left and right this can hold equally of either of the two.'

75. Cf., in opposition to this, Brentano's later identification of the concept of what is temporal with that of things in general (see n. 69 above). It remains correct, however, that even the absolute determinations of time are not individuating determinations, for arbitrarily many things can be temporally the same, indeed must be so in that they *are* at the same time.

76. Brentano is thinking here of Marty's theory of time.

77. E.g. something four-dimensional.

78. The idea is this: if, of two things standing temporally apart from each other, the one and the other were to be accepted with the same preterite mode, then it could not be plausible to us that they could not have been together in the present. The equality of mode would much rather imply the opposite: there would be lacking that differentiation which makes us find it plausible that two points in time are incompatible not only in the same but also in different things.

79. In the case of such willing however there probably comes into consideration a temporal difference between what now is and what is expected that is incomparably larger than what would correspond to the small intuitive continuum of *modi*, if, that is, we would possess at all a looking forward similar to a looking back. When I resolve to do something tomorrow, then I do not present this temporal distance intuitively, but extend in non-intuitive thinking the boundaries of my original temporal intuition, similar to the way in which we also non-intuitively conceive spatial stretches as extended beyond the boundaries of our sensory fields. If there were a special modus for the presentation of something as occurring after 24 hours, then, since a mere boundary cannot obtain in isolation, this must be a piece of modal continuum, which however was no longer connected at all to the *modus* of the present. Brentano can certainly not have intended something of this sort. It must, therefore, be a matter of a non-intuitive thinking, but it is not clear how in this case the requirement of continuous differentiation here laid down by Brentano can be met. On the way in which Brentano conceived the indeterminate presentation of

temporal distances cf. *Psychologie III*, p. 45, Eng. trans. pp. 33f.

80. Cf. *Vom Dasein Gottes*, pp. 408, 457.

81. This is naturally not to be understood as though the non-existence of something would require a cause just as would its coming to be or its preservation. The idea seems much rather to be only this, that an omniscient and morally perfect entity would not only have to know and approve what is, but also, in regard to what is not but is in itself possible, know and approve that it is not.

82. Here there follow in the dictation the words: 'It is of course remarkable that, when we form such a general presentation of spatial points, we do this by having at the same time a specifically determined presentation of some spatial continua or other, whilst here these are lacking: the general presentation of the transcendent temporal point and continuum is given to us independently'. The passage was deleted by me in the light of its later correction (e.g. in *Psychologie II*, pp. 200 and 212, Eng. trans. pp. 311f. and 320f.).

83. Cf. *Psychologie III*, p. 51, Eng. trans. p. 37.

84. Cf. *Psychologie III*, p. 6, Eng. trans. p. 6.

85. Not however as mental, but only in general, as thing. See *Vom Dasein Gottes*, p. 418, where Brentano appears to have corrected the view expressed in the *Kategorienlehre* (p.158, Eng. trans. pp. 119f.), according to which the psychic subject is grasped with evidence as zero-dimensional.

86. Cf. *Psychologie III*, pp. 4, 14, Eng. trans. pp. 4, 14.

87. The dictation bears as title only 'On the Theory of Time'. The title supplied by me reproduces its main thesis. The attempted demonstration of this thesis rests, as will be shown in the following notes, on a paralogism caused by the ambiguity of the proposition that what has happened could not be undone.

88. The proposition 'If something has once been, then it can never cease to have been' can mean either:

a. There will always be something present which stands apart from it as something later.

b. There will never be able to be something present which does not stand apart from it as something later.

Only the second of these propositions is plausible a priori, and, since of course it contains no positive assertion about something that is or will be, it would retain its validity even in the case that everything should come to be annihilated. Brentano seems not to have distinguished the two interpretations sufficiently clearly from each other. Hence he held being past, in regard to which it is to us plausible that it could never as such cease to be, to be bound up with the existence of something. The paralogism is similar to that which Brentano himself has refuted in the ontological argument for the existence of God. (*Vom Dasein Gottes*, p. 45.)

89. The second of the two is correct. What is today would have been, and would become ever further past, even after the annihilation of all things. Admittedly there would then be nothing which would stand to it as something later, but it would be impossible that anything, if it were to be, should not stand apart as something later from what is today. Only in this sense would time run on.

90. The continual uniform change of time implies only that the present points succeeding each other are continuously different. It does not however imply a change of the absolute temporal point lying in the past. It is not that what was formerly present and is now past would show its absolute specific temporal determination as changed were we in a position to present it completely. Rather, that which currently appears as in the present would manifest continuously different temporal determinations and thus also growing distances from the same past

temporal point. This would none the less become ever more past, in the sense that there could be no subsequent point in the present of which it would not be the case that every subsequent point would lie further away from it than every earlier point.

91. Just as no positive acceptance of future things is to be deduced from the a priori evident proposition that what is past can never cease to be past, so also can there be inferred no positive acceptance of a temporal change in God. There follows only the negative proposition that not even God could exist without temporal change. I emphasise this in order that no one should feel encouraged by this oversight on Brentano's part to attempt a new form of the ontological argument for the existence of God.

92. What is present does not stand apart as later from something that is merely *thought* as past; it is merely thought as standing apart therefrom.

93. C. Stumpf, *Tonpsychologie*, vol. I (Leipzig, 1883), p. 131.

94. Phenomenal space = the continuum of places that we intuit; real space = the continuum of places that we infer as existing.

95. Brentano later, in the dictations of 6 January and 21 February 1917 published in the second volume of the *Psychologie*, gave the idea of a centre of our spatial intuition yet another interpretation. According to this, there would always be some located thing that would be intuited *in modo recto*, though only in general and without qualitative determination, as something that would stand in different degrees and directions from qualified locations of the various sensory fields intuited *in obliquo*. Interpreted in this way the 'spatial centre' would acquire after all a certain analogy to the now point that is presented and accepted *in modo recto*. See *Psychologie II*, pp. 201f. and 270, Eng. trans. pp. 313f. and 363f. Cf. also the editor's note in *Psychologie III*, pp. 164f. Eng. trans. pp. 118f.

96. The series of accidental differentiation in the case of judgment, for example, would be: quality (accepting vs. rejecting), modality (assertoric vs. apodictic), and a third series, according to which judgments are differentiated into evident and blind. These intersect each other: an accepting can be evident or blind; an evident judgment can be one of acceptance or rejection; a rejecting judgment can be assertoric or apodictic; an assertoric judgment can be one of rejection or acceptance; and so on.

97. Inserted by me, since as regards what is located as such Brentano himself belongs to those who hold it to be something substantial.

98. In opposition to many earlier utterances according to which nothing at all of temporal object-differences should be found in our temporal intuition, it is here admitted that it would contain two series of relative differences, namely differences of standing apart as earlier and later, which are continuously different, and differences of present, past and future, which one could designate as the differences relative to what *is* as such. The former are object-differences, the latter modal. Still, Brentano is as before far from ascribing to us the intuition of objective temporal differences in specie. What we intuit are merely continuous temporal differences of standing apart; in other words we intuit what has become ever further removed from what is present, without there being anything that would be phenomenally altered in this itself. This is possible without contradiction, for the phenomenal equality of all succeeding present points is merely the consequence of their specific indeterminacy. Intuition is after all not, as it is according to the common opinion of the philosophers, presentation of what is individual, but of the least indeterminate universal.

99. This should read: the relative substantial differences. For temporal and local distances, too, are not accidental, but rather substantial determinations.

100. In order to prevent confusion we draw attention to the fact that what, in a

Notes

completely determinate apprehension, would appear as successively altered would be not what was formerly present and is now past, but much rather what is currently present. This would manifest continuously different temporal determinations, and for this reason also an ever-increasing distance from the same past temporal point.

101. Cf. *Kategorienlehre*, pp. 166ff., 282ff., Eng. trans. pp. 125ff., 199ff.

102. Cf. *Kategorienlehre*, pp. 182ff, Eng. trans. pp. 135ff.

103. The generic concept *coloured thing* is for example contained as 'logical part' in the species-concept *red thing*. Naturally we do not have to do here with a part-relation in the proper sense, as can be seen already in the fact that, while one can say of something red that it is something coloured, one certainly cannot say of a whole, that it is its part. Even if redness were referred to as a part of what is red, this would be only a picture. On the sense of such abstracta, cf. *Kategorienlehre*, pp. 60ff., Eng. trans. pp. 52ff.

104. Cf. *Kategorienlehre*, pp. 123, 188-96, 254, 276f., 187, Eng. trans. pp. 95f., 140-45, 182, 196f., 202f.

105. On comparative relata, cf. *Psychologie II*, p. 273, Eng. trans. pp. 365f.; further *Kategorienlehre*, pp. 236f., 242, 252, 283, Eng. trans. pp. 170f., 174, 180, 200. Contiguity will here surely have to be counted among the comparatives, for things in contact have, at the points of contact, the same determinations of time and place, even though they do not have them on the same side.

106. Cf. *Kategorienlehre*, pp. 170, 194ff., Eng. trans. pp. 127f., 143ff.

107. One can say of a material determination that it contains nothing new as compared with one that is already known, if it is included in the latter. Thus someone who knows a thing as something red learns nothing new when he hears that it is coloured. Both are material determinations, the one more specific, the other more general, the latter to be abstracted from the former. A different sense is involved however when it is said that the sentence 'B is thought by A' would contain nothing new as compared with the sentence 'A thinks B', and similarly when it is said that the sentence 'A effects B' would contain nothing new as compared with the sentence 'B is effected by A'. For here it is not the case that, of the corresponding sentences, the one is inferred from the other: rather, they say the same thing with different words. They are identical in their sense. And not both thinking and thought, effected and effecting are material determinations, but only the first word in each pair, thought thing and effecting thing being not genuine names but synsemantic.

In which sense, now, should the comparative relation posit nothing new in the two comparative objects? Brentano has elsewhere (e.g. *Kategorienlehre*, pp. 243, 252, Eng. trans. pp. 174f., 180) counted comparative determinations among the universals. In the present passage the reference to 'thought' and 'effecting' suggests a different interpretation. It seems that we would have to do here with a material predicate neither of the fundament nor of the terminus; rather, only the comparing activity itself would be a material determination, and indeed of he who engages in it. In truth, however, it would correspond more closely to Brentano's view if one says that in the comparatives both interpretations are united together. An activity of comparing is asserted *in recto*, certain more or less universal determinations of the things, determinations of a material sort being given thereby *in obliquo*.

108. This terminology seems to be inexpedient. For would not then a thinker, without there being anything in him that changed, be now relative, now quasi-relative, according to whether what he thinks exists or does not exist? Brentano thus prefers to speak of what is 'relative' wherever *modus rectus* is bound up with *modus obliquus*.

109. Read: which the speaker has as something *in the present.*

110. Certainly however what now is past had, at the time of its existence, a certain temporal determination, which I, who judges it as past, do not have. It is these absolute and transcendent temporal differences which Brentano counts as material, as contrasted with the determinations of past, present and future.

111. Namely that it belongs to a temporal continuum.

112. Brentano is thinking here of the absolute, specific temporal differences which he holds as substantial but as inaccessible to our intuition.

113. Cf. *Psychologie II*, p. 219, Eng. trans. p. 326.

114. Since there cannot exist an indeterminate temporal *modus*, though certainly temporal *modi* in general can be presented, it follows that, if we have given to us in our intuition neither absolute nor relative material temporal differences, then it must be that the presentation of a temporally more remote event is to be thought in such a way that one has the thought of a continuum of temporal *modi* which exceeds in length the intuitive continuum, and the given event is then thought as being such that it is to be presented or accepted with one of these more remote temporal *modi*. Cf. *Psychologie III*, p. 71, Eng. trans. p. 53.

115. Against this *Psychologie II*, pp. 200, 269, Eng. trans. pp. 311f., 363, *Psychologie III*, p. 71, Eng. trans. p. 53.

116. I.e. the same at the end of its duration as at the beginning.

117. I.e. neither as body nor as mind, but only in general as a thing. See *Vom Dasein Gottes*, p. 418.

118. The last two sections and probably also already sections 10. and 11. show that Brentano is concerned to resolve a difficulty which grew from the presupposition that distances cannot be intuited without the intuitive presentation of corresponding absolute positions. It seemed to him for this reason that, with the demonstration of the transcendence of absolute specific temporal determinations, there had been proved also the absence of all authentic presentation of temporal distances and thereby of *all* material temporal differences, absolute as well as relative. What suggested itself to him instead was the assumption that our so-called intuition of time is in truth a mere surrogate. And it was such a surrogate that he believed himself to have found in the modal continuum, or in the 'continuous sum of endings and beginnings'. The whole presupposition became untenable however, when Brentano finally arrived at the conclusion that not only our intuition of time but also our intuition of space is lacking in determinateness. That we none the less have intuitions of what is spatially extended, and of things that stand apart from each other therein in different degrees and manifold directions, was just as little to be denied as that we have to do here with material differences. Thus also in the last phase of his concern with the problem of time, Brentano saw himself forced to acknowledge material differences in our intuition of time, as in our intuition of space, and indeed in both cases merely relative determinations. Thus the modal continuum no longer has the task of serving as a surrogate for the transcendent material differences, which does not however lessen its significance in respect of the modification of the 'is'.

119. This does not however correspond to Brentano's own view. It is not present, past or future which he holds as specific temporal determinations (as for example red or blue are specific qualities); rather, he holds that what is presently past, must necessarily differ in its specific temporal determination from what is or will be present. The historical events which follow upon each other are distinguished from each other temporally through absolute specific temporal determinations, but every one of them enjoys in the same way having been once future, then present, then past,

so that *these* determinations could not as such count as specific.

120. After yet another dictation of 18 January 1915 *[T 18]* had called into question the subsumption of the temporal *modi* under the *modi obliqui*, Brentano here decides in its favour.

121. Sections 6. and 7. are taken from a dictation of 4 January 1915 *[T 17]*.

122. This is in contradistinction to the determinations of place, which exclude each other only in the same subject.

123. Cf. *Psychologie III*, p. 101, Eng. trans. p. 73.

124. Cf. *Psychologie II*, p. 225, Eng. trans. p. 330, where it is asserted that, in order to apprehend temporal change, an intuition of God would be required.

125. Brentano hereby surely abandons the long-held doctrine, still maintained in the dictation of 13 February 1915 ('The temporal as relative'), that our intuition of time would include merely modal differences. It is supposed now to have as its object also material differences. These do not however include absolute positions, but rather only relative differences of direction and distance.

126. What is meant is: from the place taken up by our body.

127. Here the relativity of our intuition of space is not explained as it is in *Psychologie II*, p. 201, Eng. trans. pp. 312f., according to which all parts are supposed to be intuited as standing apart from an unqualified and not absolutely specified place presented *in modo recto*. This item carries the date 21 February 1917 and is therefore Brentano's last word on this question. Cf. *Psychologie II*, p. 270, Eng. trans. pp. 363f., and *Psychologie III*, p. 164, Eng. trans. p. 118.

128. In *Psychologie III*, p.117, Eng. trans. p. 85, we read: Where Aristotle 'lists the common sensibles he does not speak of place and time, but of extension and shape, states of rest and motion, which can be the same regardless of place and time, as long as the local and temporal circumstances are the same. His view, therefore, goes hand in hand with the assumption that sensory intuition does not show absolute local and temporal differences, but only local and temporal relations.'

129. Cf. *Vom Dasein Gottes*, p. 488.

130. Conceived as a sketch of a more extensive presentation that was to have been executed by Marty's student Dr. Josef Eisenmeier. Already in the year 1873 Brentano had similarly influenced the direction of Stumpf's book *Über den psychologischen Ursprung der Raumvorstellung*.

131. Even the designation of the two theories as 'empiricist' and 'nativist' derives from Helmholtz. Cf. H. von Helmholtz, *Handbuch der physiologischen Optik* (2nd ed., Hamburg and Leipzig, 1896), pp. 608f.

132. Op. cit., p. 963.

133. On Mach's theory of space, cf. his *Erkenntnis und Irrtum* (Leipzig, 1905), pp. 331-440, Eng. trans. of 5th ed., *Knowledge and Error* (Dordrecht, 1976), 251-350 and *Die Analyse der Empfindungen und das Verhältnis des Physischen zum Psychischen* (4th ed., Jena, 1903), Eng. trans. of 5th ed., *The Analysis of Sensations*, (New York, 1959), chs. VI and VII.

134. Cf. B. Bolzano, *Paradoxien des Unendlichen*, op. cit.

135. Sections 10. and 11. are taken from a preliminary draft.

136. On substance and accident cf. the extensive investigations in Brentano's *Kategorienlehre*.

137. This had earlier also been Brentano's view. Cf. *Kategorienlehre*, pp. 109, 117f., 125, 148, Eng. trans. pp. 86, 92f., 97, 114.

138. On elasticity cf. *Kategorienlehre*, pp. 87f., 196, Eng. trans. pp. 70f., 144f.

139. Against which of course the substantiality of what is qualitative as such would stand fast from the very start, in case place should be nothing positive.

140. Add: which is clearly not a quality in the proper sense and is indeed nothing positive at all.

141. Cf. *Kategorienlehre*, pp. 296ff., Eng. trans. pp. 208ff.

142. I.e. that there would exist a place in itself in the sense that it was certainly a place, but not a determinate place.

143. From this note it is clear that Brentano, in the last phase of his investigations of the structure of our temporal intuitions, has not limited the temporal differences intuited by us to the differences of present, past and future, but that he has acknowledged also temporal object-differences and indeed as objects of primary consciousness. These are however not absolute temporal points, but merely distances in the sense of earlier and later. One should compare this with earlier utterances which manifest a certain affinity with the doctrine of Kant, in so far as time had been assigned by the latter to the sphere of inner sense. Thus in *Psychologie III*, p.52, Eng. trans. p.38, we read that in a certain sense, a sense which Kant himself was not able to make clear, 'it is true that everything that we sensorily perceive of temporal differences is perceived by inner sense and inner sense alone. For the differences that we perceive are not differences in the object, but rather differences in the way we sense external sensation, and this would not be comprehended in the absence of inner perception.' On pp. 69f., Eng. trans. pp. 51f., rest and motion are for this reason not counted among the objects of external sensation. These are utterances deriving from the years 1914 and 1915. In a dictation ('On phenomenal time') from the year 1914 *[T 43]*, it is likewise said that, because the *modi* of acceptance and the *modi* of presentation which underlie them are only objects of inner perception, it follows that time is in a certain sense to be found only in the sphere of inner sense. Only with the recognition that relative object-differences are sufficient for the intuition of what is spatial was the way made clear for the assumption of temporal object-differences.

144. Cf. *Vom Dasein Gottes*, pp. 28 and 55.

145. Marty gives the *principium identitatis indiscernibilium* an interpretation which deviates from that of Brentano. Cf. A. Marty, *Raum und Zeit*, op. cit., pp. 187ff.

146. Cf. Marty, op. cit, pp. 183f., 245f.

147. The following part of the dictation has been published already by O. Kraus in *Kant-Studien* (vol. XXV, 1920, pp. 1-23), on the basis of a not quite error-free transcription.

148. This is not expressly said in the cited book by Marty, but it was his doctrine.

149. This Brentano himself had long mistakenly believed. Only in dictations from the last years of his life is expression given to the opposite view. Cf. *Psychologie II*, pp. 200, 205, Eng. trans. pp. 312, 315; *Psychologie III*, pp. 111ff., Eng. trans. pp. 81ff.

150. B. Petronievics, *Principien der Metaphysik* (Heidelberg, 1912), vol. I, section 2, p. 109.

151. The doctrine here put forward for the sensations of sight is extended in *Psychologie II*, p. 199, Eng. trans. p. 311, to all sensory fields. (Dictation of 21 February 1917, there erroneously dated as 21 November and incorrectly amended in *Psychologie III*, p. 165, Eng. trans., p. 118.) This projection theory of Brentano has nothing to do with the theory rejected by F. Hillebrand (*Lehre von den Gesichtsempfindungen*, Vienna, 1929, pp. 112ff.) which speaks of a projecting outwards in such a way that there would have to be first of all something done by us to the colour-qualities before they would acquire a localisation as somehow at a distance. In a note to a passage in *Psychologie III*, p. 71, Eng. trans. p. 53, O. Kraus rejected also Brentano's projection- theory, though not in my opinion with sufficient

grounds. In the evaluation of this doctrine one must above all be clear as to what it is actually supposed to achieve. Does it want to explain how differences of place can be given to us in sensory intuition without it being the case that places are given to us *in specie*? For this, I believe, the doctrine is not required. For why should a psychological organisation — if such is possible at all — which would have the effect of revealing to us spatial differences of qualities among themselves without absolute determinations of place, be less probable from the start than one which made possible the intuition of the local distances of the qualities of something unqualified? The question does however arise whether Brentano's theory of projection should not be allowed to count as a good description of the psychological fact, which seems to me established, that in sensory intuition the determinations of place appear as given to us always 'from a certain standpoint' and that — in spite of the incongruence of the different sensory fields — the qualities of the different senses manifest already from the start a determinate relation to each other as though all were grouped around a centre. I would still, however, restrict the remark that qualified places are presented only *in modo obliquo* exclusively to sensations: in phantasy and conceptual thought they are surely presented *in modo recto* also.

152. Strictly speaking one cannot intuit a point for itself, but it is here after all supposed to be a matter of the presentation not of a specific place, but of something located *in genere*. Instead of 'point' one would I think better say 'some place or other'.

153. Yet still there is not lacking all qualitative determination, as in the case of the spatial centre. What we intuit as present at any given time has always some further material determination.

154. This should not be taken to imply that nothing of what is presented *in obliquo* would have an actual existence. The same thing can after all be thought at one and the same time *in obliquo* and *in recto*. Never, however, can something that is to be correctly accepted *in recto* be something merely thought.

155. This is asserted from the standpoint of the usual view, according to which matter would consist of corporeal *substances* which are sometimes at rest, sometimes in motion from place to place. Brentano however gives preference to a conception of a different sort, according to which — in place of the ether — there would exist a single unified substance at rest, the substance of space itself, and — in place of what one had earlier considered as the substance of corporeal matter — accidents which would adhere to this substance and would be carried over from one part to the other. Considered from this standpoint the demands of an empty space, i.e. of place not filled by quality, that would make motion possible, begin to acquire a discussible sense. On Brentano's theory of matter, cf. *Kategorienlehre*, pp.296ff., Eng. trans. pp. 208ff., and O. Kraus, 'Über die Missdeutungen der Relativitätstheorie' in *Naturwissenschaft und Metaphysik*, op. cit., pp.64ff.

156. Cf. *Kategorienlehre*, pp. 65, 217, 282, 285, Eng. trans. pp. 56, 157f., 200ff.

157. I.e. that unqualified place which is intuited *in recto* and is further not specified as place is in the same sense place and in the same sense substance as that place which appears to us in the visual field as coloured but is likewise not specified as place.

158. Cf. *Kategorienlehre*, pp. 51, 64, Eng. trans. pp. 46, 55.

159. Add. in a given category. For applied to determinations of different categories the name of being is, according to Aristotle, homonymous. Cf. *Kategorienlehre*, pp. 101ff., 107ff., Eng. trans. pp. 81ff., 85ff.

160. Cf. *Kategorienlehre*, pp. 25, 36, 104, 109, 127, Eng. trans. 29, 36f., 83, 86, 99.

161. An important remark, directed against attempts to derive the concept of

collective from that of a single thing or indeed to deny to the collective as such reality of any sort.

162. For Brentano's table of categories see *Kategorienlehre*, p. 405, Eng. trans. p. 148.

163. I.e. against the identification of what is (of what is real) with what is temporal in the sense of something somehow temporally determined.

Index

abstraction 5-9, 27, 41, 90, 138, 140, 145
acceleration 31, 126
acceptance, see: judgment, *modus*
accidents xxvi, 64, 92, 101, 104, 109, 114, 150, 161, 180
Agathon 94
Anaximander 111
Anaximenes 111
animals 124
anoetistic theory of space 148f.
Aquinas 51f., 100, 118, 121
Archimedes 49
Aristotle vii-x, xvi, 13, 15, 20, 30, 51, 61, 64, 78f., 90, 94f., 100, 111-113, 117f., 120f., 130, 138f., 141, 143, 150, 154f., 159, 171, 177f., 182f., 186, 193
    De anima 119, 122-23, 130
    doctrine of time 116-128
    Metaphysics 122
    Nicomachean Ethics 94, 119
    Physics 116-122
    on potentiality 34
association 71, 142f., 149
atom 6
attribute 54, 106, 158, 164
Augustine 14, 49, 51, 54, 112, 158, 177
belief 71f., 75, 101, 131
Berkeley, George 139f., 143, 165
body xxi, 2, 6f., 10, 73, 27, 32, 78, 81, 98, 105, 114, 118, 146, 150-55, 160f., 166, 169, 176, 180, 185, 195
    and boundaries 81
    substance of 82
    world of 50, 75, 139, 148
Boltzmann, Ludwig 144
Bolzano, Bernard 68f., 146, 193
Bonaventura 53, 66
boundary ix, xii, xxvi, xv-xvii, xx, 3, 5, 7, 9f., 21, 25, 50, 73, 81, 84, 89-93, 97, 109, 113, 115, 123, 129, 141, 176, 180f., 187
    coincidence of xvii, 5f., 9, 12, 108, 118, 146, 183
    connection of 11, 16
    continuous 10
    internal 10, 31, 81

running 114, 125
    space as 159
    time as 114
    see also: existence
Brouwer, L. E. J. xxii
Cantor, Georg vii, xi, 4, 146
cause 59, 65-67, 87, 120, 186, 188
    efficient 104f., 119
    and inference 73
    material 104
    'partial' cause 104
change viii, 6, 17, 23, 27, 52f., 58, 82, 84, 95, 100, 119, 127, 134, 179
    chemical, 151f.
    infinitesimal 88f., 93
    local 111
    real 87f.
    sensation of 71
    substantial 20, 151
    and time 63, 76
    uniform 13f., 20, 85, 88f., 102, 130, 135, 158, 189
chronoid 82
Chronos 13, 83, 156
Clarke, Samuel 158-162, 164f., 173, 177
collective 180
colour 7-9, 11, 21, 23, 32, 56, 110, 135, 140, 143, 147f., 152, 179, 184, 191
common sensibles 122, 137, 139, 193
comparison, act of 105, 132, 191
comparative relations, see: relations
concepts 65
    as innate 182
    origin of 1, 4-10, 55, 110, 122, 124, 143, 145
conditioning/that which conditions 65f., 77, 83, 88, 102, 126, 157, 160, 187
contact/contiguity xxii, 73, 104f., 108, 147, 183, 188, 191
continuously many vs. continuously manifold xxvi, 6f., 32-34, 36
continuum, the continuous viii, xif., xxii, 97, 145, 157
    classification of xxvi, 9-38, 78
    of differences 80
    double 46f.
    flat 28f.
    four-dimensional 115
    homogeneous 35, 45
    infinite x, 68

197

Index